Medical Implications of Basic Research in Aging

Andrew R. Mendelsohn, PhD
James W. Larrick, MD, PhD

with a Foreword by Aubrey de Grey, PhD

Eosynth Press and Regenerative Sciences Institute

© 2013 by Andrew R. Mendelsohn and James W. Larrick.

All rights reserved.

No part of this book may be reproduced in any written, electronic, recording, or photocopying without written permission of the publisher or author. The exception would be in the case of brief quotations embodied in the critical articles or reviews and pages where permission is specifically granted by the publisher or author.

Although every precaution has been taken to verify the accuracy of the information contained herein, the author and publisher assume no responsibility for any errors or omissions. No liability is assumed for damages that may result from the use of information contained within.

Cover Design: Jasmine Larrick

Interior Design: Andrew Mendelsohn and James Larrick

Publishers: Eosynth Press and Regenerative Sciences Institute

Library of Congress Catalog Number: 2013955887

ISBN: **978-0-9912162-0-8**

1. Aging 2. Longevity 3. Health 4. Science

First Edition

DEDICATION

JWL: Dedicated to my wife and life companion Jun CHEN and beautiful daughter, Jasmine.

ARM: Dedicated to my lovely wife and daughter: Katalin and Anna.

TABLE OF CONTENTS

FOREWARD.. 6
ACKNOWLEDGEMENTS.. 7
OVERVIEW AND
INTRODUCTION.. 8

Section I.   Repairing molecular damage..................................... 46

#1.   Protein homeostasis as a clinical target for increased longevity?.. 47

Section II.   Repairing intracellular substructures
              (mitochondria, telomeres)... 60

#2.   Master switch of mitochondrial biogenesis: a clinical target
      for health span enhancement? 61

#3.   Ectopic expression of telomerase safely increases health   span and
      life span.. 71

Section III.   Modulating metabolism and growth regulation... 83

#4.   Rapamycin as an antiaging therapeutic?: targeting mammalian
      target of rapamycin to treat Hutchinson-Gilford progeria and
      neurodegenerative diseases... 84

#5.   Reversing age-related decline in working memory........................ 97

#6.   Dissecting mammalian target of rapamycin to promote
      longevity.. 105

#7.   Dietary restriction: critical co-factors to separate health span from
      life span benefits.. 115

#8.	Fibroblast growth factor-21 is a promising dietary restriction mimetic.. 133
#9.	Dietary modification of the microbiome affects risk for cardiovascular disease... 144
#10.	Trade-offs between Anti-aging Dietary Supplementation and Exercise.. 155
#11	Sleep facilitates clearance of metabolites from the brain: Glymphatic function in aging and neurodegenerative diseases... 175

Section IV. Altering epigenetics, differentiation and engineering rejuvenation... 191

#12.	Overcoming the aging systemic milieu to restore neural stem cell function.. 192
#13.	Epigenetic-mediated decline in synaptic plasticity during aging... 201
#14.	The DNA methylome as a biomarker for epigenetic instability and human aging... 211
#15.	Rejuvenation of adult stem cells: is age-associated dysfunction epigenetic?... 222
#16.	Rejuvenation of aging hearts... 237

Appendix

Applied Healthspan Engineering... 245

Index.. 289

FOREWARD

In the six years since my publication of Ending Aging (New York: St. Martin's Press), numerous advances have been made in basic biomedical science that impact the possibility of controlling biological aging. Unfortunately, as I predicted, increasing understanding of the mechanisms of aging has so far had limited impact on extending longevity. Medical Implications of Basic Research in Aging provides a sampling of the most important discoveries of the past several years relevant to aging research in the context of extending human healthspan and thereby lifespan. Andrew and Jim have assembled in this volume a number of the commentaries they previously authored for Rejuvenation Research. The presentations are clearly written, jargon-free and thus accessible to those with minimal background in biology and medical science. Thorough referencing provides an opportunity for further in-depth reading, and a variety of suggestions for increasing healthspan are provided. As the appreciation spreads that aging may be brought under genuine medical control quite soon, these commentaries are precisely what the educated layman, as well as the professional biologist, needs in order to understand how those at the cutting edge of the field believe that this will come about. As Rejuvenation Research's editor-in-chief, I feel privileged to have had the opportunity to publish these articles, and I am especially grateful to Mary Ann Liebert, Inc. for participating in this means of bringing them to a new audience. I am sure that you will share my enthusiasm.

Aubrey de Grey
September 29, 2013

ACKNOWLEDGEMENTS

We appreciate the support and encourage of our colleagues. Among those most important are: Greg Adams, Mark Alfenito, Nooshin Azimi, Ramesh Baliga, Bob Balint, Carlos Barbos, Annelise Barron, Gale Bergado, Andrew Bradbury, Luis Carbonell, Casey Chan, Ben Chen, Jun Chen, George Church, David Collier, Aaco Cori, Jim Crowe, Mark de Boer, Dennis Burton, Jorge Gavilondo, Aubrey de Grey, Jeff Fang, Rafal Farjo, Elliott Fineman, Gee Fortner, John Furber, Libby Gottschalk, Julie Gottschalk, Nick Harris, Jeng Her, Eric Hoang, Sean Hu, Manley Huang, Tom Huang, Jim Huston, Miria Kaname, Matt Kerby, Brad Keller, Jin Kim, Eric Kunkel, John Lambert, Andrew Larrick, Ashley Larrick, Donald Larrick, Randy Lee, Isabelle Lehoux, Joyce Liu, Leslie Loven, Larry Lum, Jennifer Lei, Jian-xiang Ma, Julian Ma, Jim Marks, Tom Niedringhaus, Paul Parren, Andrew Perlman, Andreas Pluckthun, Jeff Price, Cary Queen, Janice Reichert, Anie Roche, Greer Rothman, Ed Schnipper, Howard Schulman, Jamie Scott, Peter Senter, Vikram Sharma, Ronnie Shaw, Mark Tepper, Ian Tomlinson, Frank Torti, Alice Tsentsiper, John Wages, Hong Wang, Jianming Wang, Shan Wang, Yuqiang Wang, Qishen Wei, Louis Weiner, Robert Whalen, Dane Wittrup, Susan C. Wright, Keith Wycoff, Geoff Yarranton, Heng Yu, Bo Yu, Gaia Zhao, Xiaoming Zhang.

The cover was kindly designed by Ms. Jasmine Larrick. Other illustrations were the work product of Wolfgang Arlo Scatman.

We especially wish to thank Susan Jensen and Mary Ann Liebert Inc. for permission to reprint our papers from Rejuvenation Research.

OVERVIEW AND INTRODUCTION

Aging Research and Healthspan Engineering

"Death comes to all."

A notable line from a very old famous book remains as true today as the day it was first written more than 2000 years ago. What is different today is the fact that many people will live long enough to experience old age before death. Everyone reading these words, it they don't succumb to some unfortunate accident or illness will grow old before they die. Yet, no one looks forward to the loss of physical vigor that accompanies aging. And rightly so, who wants to see their body and mind slowly disintegrate.

But what can be done after all? Most of us deal with this conundrum in two ways: denial before it happens and then finally during our dotage acceptance, because nothing can be done. Or can it?

The revolution in biomedical research and biotechnology over the past 35 years makes possible what was previously unthinkable: changing the substance our very biological nature. Yet, even the most optimistic futurists don't envision "curing" aging for at least 20 years. A recent US Pew poll suggests that a large majority of people are not even in favor of ending death. Even so, Google has recently (2013) sponsored the creation of a new biotech company, Calico, whose long-term goal is apparently to do just that, although there is likely as much hype as reality to their recent public relations blitz. However, the idea of reducing or even eliminating the dysfunction and illness that accompanies aging is a more popular idea and one that likely will be accomplished gradually. In other words, healthspan is more directly addressable than lifespan. Increasing healthspan has been the great success of modern medicine, as many of the diseases that have plagued human beings through the centuries can now be prevented by modern public health practices, cured with drugs or surgery or even eliminated like smallpox and polio via a worldwide vaccination campaign.

The problem of declining health in old age remains the key hurdle for further increases in healthspan. Such improvements are closely linked to our understanding of aging and its associated diseases. We have entered an era where advances in biotechnology, synthetic biology and basic biomedical research will fuel "Healthspan Engineering." We describe our preliminary thoughts on Heathspan Engineering in the Appendix to this book. Although enhanced healthspan will probably lead to increased lifespan, it is important to realize that although the processes are connected, significant lifespan enhancement is a far more difficult problem. For example, it has been estimated that eliminating cancer will only add three years on average to a person's lifespan.

To make significant progress on both healthspan and lifespan enhancement requires understanding aging, which in turn requires understanding the underlying biological processes that regulate our development from a fertilized egg to an adult, and the homeostatic processes that keep us alive. However, there is one key simplifying idea: use new parts to replace old or broken ones. The same idea which we use to maintain our cars and other mechanical and electrical devices directly applies to biological systems as well. Damaged or old tissue can be replaced by new ones. At some point in the near future we will develop the tools to enhance, reprogram and even create regenerative biological mechanisms to allow replacement of old or diseased tissue.

So the future sounds great. We can dream of curing cancer and growing new tissues and organs to replace old ones. But what can we do NOW? We have been reviewing the basic medical literature to find hints of what might one do to increase healthspan by preventing age-associated diseases or possibly even slowing fundamental aging processes. Our suggestions follow in subsequent pages, but they are couched in cautionary words and for good reason: many potentially beneficial drugs and nutraceuticals may turn out not to be, and many will have unexpected effects that may even be counterproductive. But for those of you just dying to know what they can do now, here is a sneak peak. One can likely increase healthspan by a program of moderate exercise, a smart balanced diet that avoids regular consumption of red meat, and attempting countermeasures for age-associated diseases that one is likely to suffer. The last point is the most difficult as it requires calculating personal medical risk. It is advisable that one obtain a physician's advice in making any final determination, but risk can be crudely estimated by combining personal and family medical history with inexpensive personal genomics data (from such services as 23andme) to determine which age-associated diseases one is most likely to suffer. Armed with foreknowledge of possible future diseases, it might be possible to use understanding of the underlying biological processes to select drugs and compounds found in supplements to delay or even prevent their occurrence. One extreme example of this approach are women carrying dominant mutations in the BRCA1 or BRCA2 tumor suppressor genes, who elect prophylactic surgery to remove their breasts and/or ovaries in order to ward off likely breast or ovarian cancer. Are you disappointed prospective dear reader? Were you expecting Dr. Heidegger to give you a magical rejuvenation elixir? Well, one may yet be found, but we urge caution based on the very science that promises so much.

There is a natural human tendency to breathlessly read the latest news of potential scientific or medical breakthroughs in uncritical ways. That is especially true when such news appears to promise fulfillment of a wish to live longer or healthier. But before one ends up as disappointed as Ponce de Leon was in his quest to find the Fountain of Youth, it is very useful to have a framework to place such potentially significant discoveries. Such a framework requires understanding a) how well discoveries in cells and animals translate to humans, b) that potential therapeutics may often involve tradeoffs, and c) the mechanism(s) aging works.

It is very important to understand that it is difficult to predict how well any discovery made in cell culture or in animal models will translate to benefit human beings. That's because isolated cells often behave differently than in their normal milieus and because worms, flies, mice and rats are not humans. Biologists like to focus on the impressive unity and conservation of biochemical pathways between organisms, for example mice and men are about 90% similar genetically, but there are obvious important differences. Human beings are simply better built than mice and that is why they will live on average 40 times longer. The road to fame and riches has been littered by the carcasses of biotech and pharma companies that thought they had an important therapeutic based on preclinical animal data, only to find it had limited activity in humans or even worse was harmful. If medically trained experts with years of experience can be so wrong, don't doubt for a minute that a non-expert is even on more treacherous ground. Here are some principles that we apply throughout this book. Experiments in cell culture or invertebrates need to be confirmed by experiments in animal models.

Results from studies of invertebrates or microorganisms are less reliable than mammals in predicting human responses. Although mice and rats are the most frequently used animal models they often poorly predict human responses, especially for inflammatory diseases, neurodegenerative diseases and cancer. Many published studies are more anomalous than they appear and must be independently confirmed. Human studies are better than animal studies, but even preliminary human trials, especially Phase I and II clinical trials, which tend to be small, often do not have the statistical power to predict how well a therapeutic will really work or whether an observed effect is real. Even large Phase III clinical trial data can overlook serious problems, that are sometimes discovered later, after a drug is approved by the FDA. Taken together these limitations mean that one should be very careful in applying results from basic medical research to one's own life: caveat emptor. Determination of one's comfortable level with personal risk is critical. Avoiding narrative fallacies, great sounding ideas that are untrue should always be a primary goal. For example, the idea that modulating the activity of genes identified as extending lifespan in mice will definitely extend human lifespan. Although this may be true of a few genes, many of these targets are dead ends, because humans are built better than mice and have already optimized potential gains. Further attempts at optimization could actually be counterproductive, as tradeoffs associated with altering homeostasis kick in (See **Chapter 10**).

It's been said that life is a series of tradeoffs. Not so remarkably, there are inherent tradeoffs in using almost every drug, in choice of diet and even in the amount and kind of exercise one pursues. Although this may sound almost trivial, the effects are profound and are based on the fundamental way biological systems work. For example, there are numerous studies that suggest that a daily low dose of aspirin somewhat lowers the risk of heart attack and colon cancer. However, the same mechanism that aspirin may protect against heart attacks, by blocking platelet aggregation, i.e., blood clotting, also increases the risk for internal bleeding and ulcers. Furthermore aspirin may also cause hearing loss in some people by destroying the sensory cells of the inner ear, and has been associated with Reye's syndrome in children. Current medical practice suggests that the decision to take aspirin be made by one's physician.

Why are there tradeoffs? Some tradeoffs may be due to the specificity of action of the drug or treatment. For example, aspirin has multiple molecular mechanisms of action. Aspirin inhibits prostaglandin and thromboxane synthesis through inhibition of cyclooxygenase, which in turn inhibits platelet aggregation,

By reversibly inhibiting COX-2, aspirin can uncouple oxidative phosphorylation (energy production) inside the mitochondria, with activation of AMP-activated protein kinase (AMP kinase). Tradeoffs can result from the complex linkage of biological and biochemical pathways, where affecting one component of one pathway can cause a multitude of downstream effects, not all of which are beneficial. Although not the case for aspirin, per se, some of the downstream effects of one compound may even block the beneficial effects of another as we show in Chapter 10 for the ability of Vitamin C and other antioxidants to block the beneficial effects of exercise on blood sugar and muscle metabolism. Another critical type of tradeoff is that many biochemical pathways are subject to homeostatic feedback control mechanisms, which alter the ability of a drug or treatment to affect the body. Homeostasis may modulate potentially beneficial drug-based physiological alterations. Sometimes this can make the drug ineffective over time, other times this activity may augment the danger of withdrawal of the drug. Aspirin provides an example of the latter: ending longterm use of aspirin is associated with an increased short-term risk of heart attack due to increased cyclooxygenase expression and clotting resulting from the body's attempt to compensate.

Understanding the basis of aging helps to place the potential benefits of any treatment designed to impede or reverse aging in perspective. Because aging results from the decay in homeostasis and functional reserve of normal physiological and biochemical processes in an organism the transmission of pain information to the brain, modulation of the body's internal thermostat in the hypothalamus, and inflammation.

Over time, it is likely that a complete description of the mechanisms by which humans age will require a complete description of how humans function at the molecular, cellular, organ, and systemic levels. However, we do understand enough today to potentially intercede and to choose productive directions for future research.

So first the big question: Why do we age? The answer is that we (and all other life on earth) have been selected by evolution to accomplish one great task: reproduction. Our body is just a supporting player. It's a well-built supporting player with the capacity to live as long as 120 years in some cases, but it still remains a supporting player for the cells that really count: the germ cells (sperm and egg cells). These cells have a degree of immortality in that they can potentially produce a new man or woman, who in turn can produce offspring and so on. However, the rest of the body, including our brains which encode our precious consciousness is made of cells that have not been selected to endure. Evolutionary selection to maintain a sexually mature individual diminishes with time as the probability of successful reproduction increases.

The second big question is: How do we age? This is the subject of a great amount of work and speculation. It is our opinion that the most useful way to look at aging is that adult human beings (and all but a few animals) completely lack rejuvenative homeostatic maintenance mechanisms. In a typical animal's life, a developmental program is executed, when it is finished the organism is dependent on pre-existing homeostatic mechanisms. At the cellular level, in multicellular animals these mechanisms require maintaining cell differentiation state, specialized cell function, and eliminating any damage from biochemical processes and the environment. That damage to cellular biomolecules, organelles, and the cells themselves occurs over time is well understood, and precise characterization of that damage is ongoing. What is usually ignored is the question of what is absent that allows such damage to accumulate, since we know that germ cells at least are capable of a kind of indirect immortality. As a crude example consider the moth born without a mouth, in this case we know exactly what is missing to cause the "damage," although here the very idea of "damage" clearly misses the mark. We hypothesize that absence of critical homeostatic mechanisms is the key to aging, and among those mechanisms that are absent, cell and tissue replacement (regeneration) is the one most often critically absent. Furthermore, besides direct biomolecular damage, the intrinsic strength of the underlying biochemical networks to stochastic noise probably plays an important role in the loss of cellular and tissue function observed with aging, especially that relating to changes in epigenetic state and cell differentiation.

So why does this matter? Because it points to ways in which diseases of aging might occur and to why simplistic interventions that target one type of damage, regulatory molecule, or epigenetic change such as enhancing telomere length (Chapter 3) are unlikely to be as significant as they may first appear. Correctly understanding how aging occurs is quite important in interpreting the implications of basic research results.

We have organized reviews of basic research into 4 sections that correspond to the basic processes that can be altered to affect aging:

- Repairing molecular damage

- Repairing or rejuvenating cellular substructures

- Modulating metabolism and growth regulation, and

- Rejuvenation through modifying epigenetics and cell differentiation.

Please note the conspicuous absence of enhanced regeneration, which we earlier indicated to be perhaps the most promising means to achieve a breakthrough in health and lifespan. Unfortunately, safely engineering new regenerative processes has not yet been reported and will require significantly more sophisticated biotechnology than is now attempted clinically.

Aging models are of more than academic interest, because they have profound impact not only on the research community, but also on the public. There is a cottage industry of aging models in the scientific literature. Like the blind men of legend describing an elephant each tends to focus on one aspect of aging, often coming to radically different conclusions. Eventually a consensus will be reached in which the correct aspects of each model will be unified and an "elephant" of a theory might emerge.

The best example of a highly influential damage theory of aging known as Harmon's free-radical theory of aging, which hypothesizes that metabolic activity leads to generation of harmful reactive oxygen species that then damage biomolecules, cells and beyond leading to aging and death. This theory was extended to the mitochondrial free radical theory of aging, which localizes the primary source of damaging free radical production to the mitochondria. Data has accumulated to show that reduction of oxidative stress is associated with prolongation of life expectancy in animals, especially invertebrates. These theories logically led to the idea that antioxidants might be able to prolong lifespan and healthspan. This idea has become very widespread and popular both in and outside of scientific circles. However, most studies have found a lack of correlation between antioxidants and human health benefits. Some reports even suggest that antioxidants in humans may promote cancer or increase incidence of diseases associated with detrimental effects on longevity. One very interesting effect of antioxidants is that they appear to block the beneficial effects of exercise in humans, in some cases contrary to their effects in mice (see **Chapter 10** for an extended discussion).

Updates and non-technical summaries of the chapters

#1. Protein homeostasis as a clinical target for increased longevity?

There are various kinds of biomolecular damage observed with aging. Damage to DNA, RNA, lipids and proteins have been reported. Experiments to explore ramifications of various damage theories of aging account for the largest body of aging research. One of the more serious changes that occurs is inappropriate aggregation of proteins. Although damage via free radicals is considered the usual culprit, inappropriate modification and fundamental stickiness and metastability of the relevant proteins probably plays a significant role. The accumulation of damaged proteins is considered to be a major mechanism by which homeostasis declines during aging. In this chapter, we consider some ways to overcome an age-associated decline in protein homeostasis mechanisms, especially those that maintain protein quality control which contribute to aging. Reducing protein aggregation or stimulating aggregate removal would have profound implications for the treatment of numerous degenerative pathologies, including Alzheimer's, Huntington's, and Parkinson's diseases, in which dysfunctional protein homeostasis plays a significant role. However, the role protein aggregation plays in aging needs to be better quantified.

The ability of several compounds to block toxicity of an aggregatable human amyloid protein fragment (associated with Alzheimer's disease) or polyglutamine protein (associated with Huntington's disease) in an engineered nematode was investigated by Alavez and colleagues. They found that the flavanoid thioflavin T, curcumin, and the antibiotic rifampicin blocked the ability of aggregated amyloid to kill the worms. These compounds were known to disaggregate amyloid, likely through the biophysical stabilization of the protein, but the authors of the study found that activity of protein quality control pathways were also essential for these compounds to inhibit toxicity.

The mechanisms underlying protein homeostasis present attractive targets for development of pharmaceutical "proteostabilins", drugs that stabilize aggregated proteins and stimulate protein quality control pathways to extend health- and lifespan. The biggest problem here is that these studies were performed in nematodes, so their ultimate application to human health remains uncertain. Curcumin is readily available in turmeric and in supplements and appears to be relatively harmless at available doses, so people who have an enhanced probability of Alzheimer's disease (e.g. APOE4 homozygotes) may want to experiment. However, a small clinical trial of curcumin to treat 36 Alzheimer's patients in 2012 failed to show any benefit. These results do no address the possibility of disease prevention. Unfortunately, very few studies to determine whether a compound can prevent Alzheimer's in humans have been performed. The science needed to recommend curcumin to prevent Alzheimer's does not yet exist.

The development of better targeted anti-aggregation drugs promises a new class of anti-aging therapeutics. However, controlled protein aggregation is the basis of much of normal cell signaling. The identification of appropriately targeted anti-aggregation agents that have desirable biological effects may not be as easy as we would like.

#2. Master switch of mitochondrial biogenesis: a clinical target for health span enhancement?

It has been known for some time that most animal somatic cells lack the enzyme telomerase, which limits the possible number of cell proliferations in a way directly related to the length of the telomeres. Telomeres are necessary for proper linear DNA chromosomal replication. In the absence of telomerase, telomeres gradually shorten with each round of DNA replication. The complete absence of telomeres would prevent the full DNA replication of chromosomes, but short telomeres induce cell (replicative) senescence which in turn blocks further cell proliferation before a cell's telomere reserve is completely exhausted. Senescence can also be triggered by other effectors, including various types of cell stress. The presence of senescent cells in mice has been correlated with aging-associated dysfunction.

In an important paper, Sahin *et al.* connect replicative senescence with mitochondrial dysfunction. They suggest that the master regulators of mitochondrial biogenesis PGC-1-alpha and PGC-1-beta play an important role in aging-related pathologies. They found mitochondrial dysfunction in mice which had been engineered to have short telomeres by knocking out either of two critical telomerase subunits. This dysfunction occurred in non-proliferating cells such as the liver and heart. They discovered that mitochondrial master regulators PGC-1-alpha and PGC-1-beta were down-regulated through a p53 dependent mechanism, and that liver-targeted gene therapy with PGC-1-alpha and PGC-1-beta restored liver function.

The connection between telomere shortening and mitochondrial biogenesis was established in an artificial mouse model, which suffers from short telomeres due to the genetic knockout of telomerase. Direct connection with human health and aging is uncertain, as the number and role of senescent cells, including those induced by stress or replication is unknown. However, it is known that mitochondrial biogenesis is often reduced in elderly humans, indicating that stimulation of PGC-1-alpha and PGC-1-beta will likely benefit aged people.

AMPK, Sirt1 and exercise are all known to increase PGC-1-alpha and PGC-1-beta and stimulate mitochondrial biogenesis. Given the potential benefits of augmented PGC-1-alpha and PGC-1-beta function, new drugs to increase expression or activity of these proteins are expected to have clinical benefit. Several existent drugs or nutraceuticals known to increase the activity of AMPK or SIRT1 also stimulate PGC-1 proteins, at least indirectly. For example, metformin is known to stimulate AMPK via inhibition of AMP deaminase, and multiple studies indicate stimulation of PGC-1. Moderate exercise may be the safest means to increase PGC-1. Although high-intensity interval training raises both SIRT1 and PGC-1-alpha levels, low-volume sprint interval exercise in cyclists appears to achieve similar benefits.

#3. Ectopic expression of telomerase safely increases health span and life span.

It is well known that telomeres of adult human cells that are not involved with reproduction shorten with each cell division as a result of a lack of the enzyme telomerase, which can extend telomeres. In other words, telomeres grow shorter with the age of an organism at least in cells that divide. Increasing telomere length to slow or prevent aging is a popular idea, especially in some segments of the lay community, but the science is controversial. The absence of telomerase from somatic cells of mammals has significant consequences for aging. First, it limits the number of potential cell divisions and in so doing sets limits on both life span and cancer cell proliferation. Second, shortened telomeres are known to result in physiological dysfunction, including playing a role in human diseases such as Werner syndrome and ataxia telangiectasia. Ectopic expression of the catalytic subunit of telomerase, telomerase reverse transcriptase (TERT), has been reported to extend life span by as much as 40% in cancer-resistant mice. On the other hand, ectopic expression of TERT promotes cancer in normal mice.

However, transient induction of TERT by an astragalus-derived compound, which is available to the public as an unregulated nutritional supplement, increases health span without an apparent increase in cancer incidence in mice. In an attempt to confirm and extend these studies, a study using a method which induces permanent changes in mice was performed. Ectopic expression of TERT using adeno-associated virus serotype 9 (AAV9)-based gene therapy in adult mice increases both health span and life span without increasing cancer incidence. Available evidence suggests that increases in life span may require both elongated telomeres and the continuous presence of telomerase to stimulate the WNT/beta-catenin signaling pathway. The recent observation that WNT/beta-catenin signaling can stimulate TERT expression raises the possibility of a positive feedback loop between TERT and WNT/beta-catenin. Such a positive feedback loop implies that safety must be carefully considered in the development of drugs that stimulate telomerase activity, because increased WNT/beta-catenin signaling is a hallmark of many cancers. Studies of this scale in mice just can not rule out real cancer risk to humans (or mice for that matter).

It is probably not safe to generalize from these mouse studies to humans. Further Investigation of the role telomere length plays in human aging Is warranted. Although there is evidence suggesting that increased telomere length correlates with longevity in humans, there is also evidence from genome-wide association studies that mutations that result in increased telomere length associate with an increased risk of multiple myeloma and colon cancer. There may very well be a potential tradeoff in attempting to lengthen one's telomeres. Clever folks on select internet forums have apparently decided to take the risks associated with increasing their telomerase length, but hope to hedge against problems by also attempting to increase expression of tumor suppressor genes. Unfortunately, without clinical trial data, or the possibility of quantitative systems simulations of human biochemistry and physiology which don't yet exist, it is impossible to know whether their approach is efficacious or safe.

#4. Rapamycin as an anti-aging therapeutic?: targeting mammalian target of rapamycin to treat Hutchinson-Gilford progeria and neurodegenerative diseases.

Mammalian target of rapamycin (mTOR), a serine/threonine kinase and component of the mTORC1 signaling complex, acts as an energy, nutrient, growth factor, stress, and redox sensor to increase protein synthesis and decrease macroautophagy. mTORC1 plays a central role in the maintenance of cellular homeostasis and deterioration of its control is seen in aging. The Food and Drug Administration (FDA)-approved immunosuppressive macrolide rapamycin binds immunophilin FKBP12 (FK506-binding protein) to inhibit mTORC1. <u>Unlike most other interventions tested to date, inhibition of mTORC1 by rapamycin extends life span in old mice,</u> likely by a combination of increased autophagy and decreased mRNA translation. Some have promoted rapamcyin as a potential poor man's "fountain of youth," and no doubt there are some who are experimenting with this drug in hopes of reaping anti-aging benefits (see **#6** for more).

Rapamycin does not only increase lifespan in old mice, but also appears to treat a mouse model of premature aging in humans: Hutchinson–Gilford progeria syndrome.

Hutchinson–Gilford progeria syndrome (HGPS) is a tragic, lethal genetic disorder affecting children that is characterized by symptoms of premature aging, such as atherosclerosis.

In mice engineered to express a mutant human protein, progerin, that causes HGPS, rapamycin increases autophagy, a process by which cells devour themselves to house clean,, and thereby reduces accumulation of progerin, an alternate spliced form of lamin A/C, that forms insoluble toxic aggregates. Rapamycin treatment results in reduced HGPS-associated nuclear blebbing, growth inhibition, epigenetic dysregulation, and genomic instability. Rapamycin-induced autophagy also suppresses symptoms in mouse models of Alzheimer's, Parkinson's, and Huntington's diseases, where toxic insoluble protein aggregates accumulate. On the basis of these results, modulation of mTORC1 function is a promising target for the development of therapeutics for neurodegenerative diseases and HGPS. Rapamycin is the obvious candidate for near-term evaluation in the treatment of these diseases.

However, the substantial set of rapamycin-associated adverse effects, as well as the lack of aging-specific human data, should caution the routine use of rapamycin as a general-purpose anti-aging agent. Further caution should be taken based on the results of a recent study showing that much of the increase in longevity conferred by rapamycin in mice was due to its anti-tumor activity. The use of safer, but perhaps weaker, indirect mTORC1 inhibitors, such as metformin and resveratrol, may prove useful. Further study will ascertain whether such compounds extend human health or life span.

#5. Reversing age-related decline in working memory.

Higher cognitive functions, such as working memory and the ability to focus attention, decline as people age. Recently, it was reported that decline in working memory in aging rhesus monkeys correlates with the loss of activity of a specific set of neurons in the prefrontal cortex. The activity of these neurons can be rescued by stimulating alpha-2 adrenergic receptors, inhibiting cyclic adenosine monophosphate (cAMP) signaling, or closing potassium channels that are known to inhibit firing and synaptic connectivity. Agents that stimulate neurons expressing alpha-2 adrenergic receptors may prove useful in treating working memory loss in humans.

The use of alpha-2A receptor agonists, such as guanfacine, a Food and Drug Administration (FDA)–approved drug for attention-deficit disorder may prove useful for elderly patients with working memory dysfunction, and a clinical trial is underway to test this hypothesis. Positive results in this clinical trial may open the door to treating working memory in humans.

#6. Dissecting mammalian target of rapamycin to promote longevity.

Treatment with rapamycin, an inhibitor of mammalian target of rapamycin complex 1 (mTORC1) can increase mammalian life span at least in mice. However, extended treatment with rapamycin results in increased hepatic gluconeogenesis concomitant with glucose and insulin insensitivity through inhibition of mTOR complex 2 (mTORC2). <u>In other words extended treatment with rapamycin increases risk for diabetes type II.</u>

Is there a way around this potential problem with rapamycin? Genetic studies show that increased life span associated with mTORC1 inhibition can be at least partially decoupled from increased gluconeogenesis associated with mTORC2 inhibition. Adenosine monophosphate kinase (AMPK) agonists such as metformin, which inhibits gluconeogenesis by down-regulating expression of glucose-6-phosphatase and phosphoenolpyruvate carboxykinase, might be expected to block the glucose dysmetabolism mediated by rapamycin. The search for inhibitors of the mTORC1 component Raptor may prove a productive approach to create a better mTOR inhibitor. So rapamycin itself has a significant drawback for healthspan enhancement, but derivatives of rapamycin may be engineered to avoid the diabetes pitfall.

As previously mentioned, important recent work suggests

that the ability of rapamycin to increase longevity may largely be due to its anti-tumor activity, and thus it may confer no benefit to intrinsic aging

#7. Dietary restriction: critical co-factors to separate health span from life span benefits.

The most hyped potential anti-aging regimen involves reduction of total energy intake by adhering to a very strict diet. Dietary restriction (DR), a.k.a. caloric restriction, typically a 20%–40% reduction in "normal" nutritional energy intake, has been reported to extend life span in diverse organisms, including yeast, nematodes, spiders, fruit flies, mice, rats, and rhesus monkeys. When in works, DR works best when began youth, but it is still beneficial even in adults. Interestingly, the magnitude of the life span enhancement appears to diminish with increasing organismal complexity. However, the extent of life span extension has been notoriously inconsistent, especially in higher mammals.

Recently, Mattison *et al.* reported that DR does not extend life span in rhesus monkeys in contrast to earlier work of Colman *et al.* Examination of these papers identifies multiple potential confounding factors. Among these are the varied genetic backgrounds and composition of the "normal" and DR diets. In monkeys, the correlation of DR with increased health span is stronger than that seen with life span and indeed may be separable.

Recent mechanistic studies in fruit flies implicate non-genetic co-factors such as level of physical activity and muscular fatty acid metabolism in the benefits of DR. These results should be followed up in mammals. Perhaps levels of physical activity among the cohorts of rhesus monkeys contribute to inconsistent DR effects. To understand the maximum potential benefits from DR requires differentiating fundamental effects on aging at the cellular and molecular levels from suppression of age-associated diseases, such as cancer. To that end, it is important that investigators carefully evaluate the effects of DR on biomarkers of molecular aging, such as mutation rate and epigenomic alterations. Several short-term studies show that humans may benefit from DR in as little as 6 months, by achieving lowered fasting insulin levels and improved cardiovascular health. Optimized health span engineering will require a much deeper understanding of DR.

The bottom line is that DR studies show a potential divergence between lifespan and healthspan in humans and monkeys. DR studies in humans have involved relatively small numbers of subjects. However, there are dedicated groups of people who voluntarily restrict calories for potential health benefit. Larger studies are needed to identify possible DR tradeoffs such as reduced muscle mass, libido, etc.

DR as a health and lifespan enhancing regimen remains controversial. Because only a select group of individuals appear to have the will power to adhere to the diet under threat of constant hunger, it is possible that DR will never be popular even if it is proven to work. On the other hand, if humans are really better built than mice, and we are, then the various improvements already made by evolutionary processes may significantly reduce the benefits of DR, especially regarding lifespan enhancement. An effort to precisely identify how the various biochemical differences between humans and other organisms, especially mice will be of great benefit in identifying potential ways to extend health- and lifespan.

#8. Fibroblast growth factor-21 is a promising dietary restriction mimetic.

Dietary or caloric restriction (DR or CR), typically a 30%–40% reduction in ad libitum or "normal" nutritional energy levels, has been reported to extend life span and health span in diverse organisms, including mammals. Although the life span benefit of DR in primates and humans is unproven (see **#7**), preliminary evidence suggests that DR confers health span benefits. A serious effort is underway to discover or engineer DR mimetics. The most straightforward path to a DR mimetic requires a detailed understanding of the molecular mechanisms that underlie DR and related life span–enhancing protocols. In this chapter we discuss one potential DR mimetic.

Increased expression of fibroblast growth factor-21 (FGF21), a putative mammalian starvation master regulator, promotes many of the same beneficial physiological changes seen in DR animals, including decreased glucose levels, increased insulin sensitivity, and improved fatty acid/lipid profiles. Ectopic over-expression of FGF21 in transgenic mice (FGF21-Tg) extends life span to a similar extent as DR in a recent study. FGF21 may achieve these effects by attenuating growth hormone (GH)/insulin-like growth factor-1 (IGF1) signaling.

Although FGF21 expression does not increase during DR, and therefore is unlikely to mediate DR, it does increase during short-term starvation in rodents, which is a critical component of alternate day fasting, a DR-like protocol that also increases life span and health span in mammals. Various drugs have been reported to induce FGF21, including peroxisome proliferator-activated receptor-a (PPARa) agonists such as fenofibrate, the histone deacetylase inhibitor sodium butyrate, and adenosine monophosphate (AMP) kinase activators metformin and 5-amino-1-b-D- ribofuranosyl-imidazole-4-carboxamide (AICAR). Of these, only metformin has been reported to extend life span in mammals, and the extent of benefit is less than that seen with ectopic FGF21 expression I mice. Perhaps the most parsimonious explanation is that high, possibly un-physiological, levels of FGF21 are needed to achieve maximum life span and health span benefits and that sufficiently high levels are not achieved by the identified FGF21 inducers. More in-depth studies of the effects of FGF21 and its inducers on longevity and health span are warranted.

Whether FGF21 or inducers of FGF21 directly lead to anti-aging therapeutics, the search for DR mimetics continues, and it is even possible that someone will stumble on a compound or combination of compounds that increases human health and lifespan even more than DR. However, the better we understand how human biochemistry translates into higher level physiology, the greater the chance that a real anti-aging therapy can be developed.

#9. Dietary modification of the microbiome affects risk for cardiovascular disease.

The incidence of cardiovascular disease (CVD) increases with age and is associated with some syndromes that exhibit aspects of premature aging, such as progeria. Various factors are thought to contribute to the progression of CVD, including hypertension, hypercholesterolemia, diets rich in saturated and trans fats, etc. Recent reports have uncovered an important connection between diet, the microbiome, and CVD. Dietary carnitine (present predominately in red meat) and lecithin (phosphatidyl choline) are shown to be metabolized by gut microbes to trimethylamine (TMA), which in turn is metabolized by liver flavin monoxygenases (especially FMO3 and FMO1) to form trimethylamine-N-oxide (TMAO). High levels of TMAO in the blood strongly correlate with CVD and associated acute clinical events. Plasma TMAO levels may be an important clinical biomarker for CVD. The data suggest that that presence of specific as yet unidentified microorganisms in the gut linked to diet are required for high TMAO levels and TMAO-mediated CVD progression. Development of novel therapeutic approaches to manipulate gut flora may help treat CVD.

For those at risk of CVD, which may include most human beings, control of TMAO levels may prove quite prudent. Of immediate relevance is that supplementation with choline or L-carnitine may be unwise in most circumstances, despite possible benefits in cognitive or muscle function, respectively, unless TMAO levels are carefully monitored. On the other hand, it well known that choline is an important nutrient. Low levels of choline can lead to organ dysfunction. Choline is necessary for the production of the neurotransmitter acetylcholine and PC is a critical component of cell membranes. Complete elimination of choline or PC from one's diet would be unproductive as well as difficult to achieve. However excess choline, such as that found in eggs, may be worth avoiding. <u>These studies were conducted in humans and given that the data clearly support this mechanism of action should be taken very seriously.</u> The various warnings about eating too much red meat that have circulated in recent years, has new scientific backing, with a completely unexpected mechanism. Such warnings probably should be heeded.

#10. Trade-offs between Anti-aging Dietary Supplementation and Exercise.

In otherwise healthy adults, moderate aerobic exercise extends lifespan and likely healthspan by 2-6 years. Exercise improves blood sugar regulation, and resistance exercise increases or maintains muscle mass, and is associated with improved cognitive function. On the other hand, evidence for antioxidant supplements increasing longevity in humans is lacking. On the contrary, transient hormetic increases in ROS, for example associated with exercise, are actually associated with increased mammalian healthspan and lifespan.

Recent studies in humans suggest that antioxidants such as vitamins C, E, resveratrol, and acetyl-N-cysteine blunt the beneficial effects of exercise on glucose sensitivity and blood sugar regulation, likely through direct inhibition of ROS signaling. Alternately, other studies suggest that vitamin C has beneficial effects on exercise-associated dysfunction: inhibiting exercise-induced bronchoconstriction. These data suggest that there are tradeoffs between potential benefits and harm from antioxidant dietary supplementation. Specific biomolecular interactions for each antioxidant also will be important.

Omega-3 (n-3) polyunsaturated fatty acids (PUFAs) have anti-inflammatory activity that is not mediated through direct ROS inhibition. Although data is limited in humans, n-3 PUFAs do not seem to blunt blood sugar regulatory benefits of aerobic exercise, and actually increase anabolic activity in skeletal muscle. However, another kind of tradeoff may exist with PUFAs, at least for men: a recent large clinical trial demonstrates an association of omega-3 fatty acids blood levels with increased incidence of prostate cancer, especially aggressive prostate cancer. Together these results suggest that there are significant tradeoffs in the use of dietary supplementation for prevention and treatment of diseases associated with aging. In the case of Omega-3 fatty acids the risks may be gender specific: so far there is no increased risk for women, only potentially men.

Such tradeoffs may result from underlying intertwined homeostatic mechanisms. For most individuals, moderate exercise is of significant benefit. Careful attention to individual and family medical history, and personal genomic data may prove essential to make wise dietary and supplement choices to be combined with exercise.

As we explained above, tradeoffs inherent in the response of complex biological organisms to agents complicate healthspan-promoting strategies. Personalizing sets of treatments may be the best way forward to avoid tradeoff-associated pitfalls.

#11. Sleep facilitates clearance of metabolites from the brain: Glymphatic function in aging and neurodegenerative diseases.

Ben Franklin's "Early to bed and early to rise, makes a man healthy, wealthy and wise" encapsulated the wisdom of getting a good night's sleep that was known to Hippocrates and probably to prehistory. Yet we haven't really understood the biology of why we need a good night's sleep.

It's not just humans that need to sleep. For mice, rats, and flies we know that lack of sleep kills. Although, humans are far better built than these animals, and can withstand significant amounts of sleeplessness, there are disorders such as fatal familial insomnia, where lack of sleep may indeed kill, although the underlying pathology is driven by an inherited prion-like process (think Mad Cow Disease). The disease is probably telling us that even more than simple sleep is necessary, because victims can actually reach the early stages of non-REM (rapid eye movement) sleep-- what they can't do is reach deeper sleeping states, including REM sleep where dreams occur. "To sleep perchance to dream" doesn't apply.

The idea that sleep acts to clear waste is not a new one. However, recently the glymphatic system has been discovered that clears the brain of waste products to protect fragile neurons from the toxic side-effects of metabolism. It works in an analogous way to the lymphatic system (**Figure 11.1**). Unlike most of the other organs in the body, the brain lacks lymphatic vessels for clearing excessive interstitial fluid, metabolic waste products and unneeded extracellular biomolecules, such as amyloid beta, tau, and alpha-synuclein that are linked to neurodegenerative diseases. However, the brain instead has the glymphatic system which accomplishes the same purpose.

The glymphatic system uses convective flow to clear biomolecules. Cerebrospinal fluid (CSF) enters the brain parenchyma (functional parts) along arteries and exchanges with the interstitial fluid (ISF) that bathes the space between brain cells. The ISF carries extracellular biomolecules from the interstitial (extracellular) space in the brain to drain through the veins. This system was termed the 'glymphatic' pathway due to its dependence upon glial cells and its performance of peripheral 'lymphatic' functions in the CNS (**Figure 11.1**). How important is it? From initial reports it became clear that the glymphatic system was responsible for clearing 65% of Abeta, one of two key proteins thought to cause Alzheimer's disease, and may be **the** major pathway by which metabolites are removed from the extracellular spaces of the brain.

It turns out that sleep drives the brain to stimulate the glymphatic system into house-cleaning mode, by causing the astrocytes, cells that support nerve cell function, to shrink, thereby increasing the brain's extracellular space and allowing the CFS to flow into the brain's inner most reaches to interact with the IFS (**Figure 11.2**). Anesthesia can achieve a similar effect. Moreover, similar effects are obtained by blocking adrenergic signaling, which normally helps trigger and maintain wakefulness – adrenaline (epinephrine) is representative of a positive adrenergic signaling molecule.

Work on the glymphatic system has been performed primarily in rodents, but there are good reasons that such a general system would be evolutionary conserved in mammals, if not vertebrates and possibly beyond.

If the glymphatic system is really important in keeping the brain clear of proteins that can aggregate into toxic precipitates, then we would expect to see correlations of sleep with neurodegenerative diseases that are driven by the accumulation of protein-rich plaques and tangles. Evidence is accumulating that this is indeed the case.

Of the many reported correlations of sleep with neurodegenerative diseases, one clinical study suggests that duration and quality of sleep may predict the occurrence of Alzheimer's disease (AD). Even more significant is that sufficiently long high quality sleep appears to protect a set of subjects who are prone to AD because they carry the ε4 allele of the APOE gene. People with on copy of APOE4 ε4 are 2-4 fold more likely to suffer AD by age 79, and people who have two copies are 10 times more likely (70% chance of AD by 85). Quality sleep reduced the risk of these people getting AD by 33% and reduces the number of plaques and tangles associated with AD, suggesting that appropriate sleep may delay or prevent AD. We believe if such data are confirmed in subsequent work, a public health initiative should be engaged to help ensure that middle to older aged people get better sleep.

What is the best way to get better sleep? Apparently not sleeping pills, but rather moderate exercise during the day and a 30 minute nap between 1 and 3pm in the afternoon. Napping has been associated with preventing AD in a small Japanese study. Napping must be limited to around 30 minutes, as longer naps (> 1 hour) are actually counterproductive.

So should you sleep you life away to avoid AD? Probably not. Mortality studies show that too much sleep is associated with an even greater mortality rate than too little. Here too there is a Goldilocks effect: you need just the right amount of quality sleep. Furthermore, it appears that for humans the timing and depth of sleep is important. It will be very interesting to see if there are subtleties in the workings of the glymphatic system, or in the other clearance mechanisms (there are receptors on the surface of brain cells that help clear some of the toxic proteins) that correlate with sleep state in humans.

#12. Overcoming the aging systemic milieu to restore neural stem cell function.

In recent years, biologists discovered that new nerve cells are constantly being generated (neurogenesis) in two regions of our brains. As mammals age, the rate of neurogenesis in the brain declines with a concomitant reduction in cognitive ability. Recent data suggest that blood-borne factors are responsible for inhibition of neurogenesis.

When the circulatory systems of old and young mice are connected experimentally, the old mice experience increased neurogenesis and the young mice exhibit less neurogenesis, suggesting the importance of systemic circulating factors. Chemokine CCL11/eotaxin has been identified as a factor that increases with aging. Injections of CCL11 inhibit neurogenesis in young mice, an effect likely mediated by CCR3 receptors on neural stem cells. Identification of a specific factor that plays a causative role in stem cell dysfunction in aging is consistent with data showing that transforming growth factor-b (TGF-beta) inhibits satellite (muscle stem) cell-mediated repair. Together, these data suggest that the systemic milieu plays a critical role in the aging of adult stem cells. Because adult stem cells help maintain homeostasis by providing the possibility of replacing metabolically damaged differentiated cells, aging of the systemic milieu and stem cell niches may drive functional decline during aging. The identification of a specific systemic change suggests that aging is more amenable to therapeutic modulation than work on global metabolism-derived damage and cellular senescence implies.

Although neurogenesis plays a role in memory and cognitive function in mice, it is only part of a more complex set of mechanisms to ensure plasticity. There are other aspects of brain function that decline during aging that may be addressed by potential intervention. To the extent that neurogenesis is a potential target to improve age-related decline in cognition, there are several reported means to achieve increased neurogenesis: Supplementation with curcumin and apigenin, exercise, and dietary restriction. The evidence that these may be helpful in humans comes only from animal models and must be considered quite preliminary.

Targeting CCL11 function must be considered completely speculative at this time, because it may be necessary to inhibit the function of several plasma-borne factors to effectively induce neurogenesis in older people. Substances that have been reported to inhibit CCL11 function include nobiletin and 7,4'-dihydroxy flavone from *Glycyrrhiza uralensis*. However, there is no available evidence that these molecules will induce neurogenesis, although it is possible that they would be able to achieve some beneficial effect, especially in conjunction with treatments that do appear to stimulate neurogenesis at least in culture and mice.

This work provides hope that incremental anti-aging therapies may be possible to maintain critical cell function by countering age-associated changes in the extracellular milieus in our bodies.

#13. Epigenetic-mediated decline in synaptic plasticity during aging.

Cognitive decline observed in aging mammals is associated with decreased long-term synaptic plasticity, especially long-term potentiation (LTP). Synaptic plasticity is of great importance to brain function: learning and memory are thought to result from specific alterations in the strength of synaptic connections. The ability to alter the strength of synaptic connections is called synaptic plasticity.

Recent work has uncovered a connection between LTP,

histone acetylation, and brain-derived neurotrophic factor (BDNF)/neurotrophin receptor B (trkB) signaling. LTP, histone acetylation, and BDNF/trkB signaling decrease in old animals, Because an apparent positive feedback loop links these processes, treatment with histone deacetylase inhibitors or a trkB agonist restores LTP in the hippocampus of old animals. These results coupled with exciting work on histone methylation and life span in *Caenorhabditis elegans* suggest that epigenetic changes may play a significant role in aging. Such dysfunctional epigenetic pathways may provide novel targets for cognitive enhancing therapeutics.

The use of global HDAC inhibitors to promote memory in the elderly is a possibility. For example, the HDAC class 1 The BDNF agonist 7,8-DHF is a promising compound for treatment of defective cognition as well as for seizure, stroke, and Parkinson disease. 7,8-DHF is found in low doses in fruits and vegetables, although levels are unlikely to be in the therapeutic range. Although the pharmacokinetics and toxicology have not been studied, like other flavonoids they are expected to be quite safe. We predict that this compound will be available as a nutraceutical should favorable preclinical results continue to be reported. Deoxygedunin is another naturally derived BDNF agonist with potent neurotrophic activity that may have similar beneficial effects on cognition.

It should be noted that exercise has been reported to stimulate BDNF levels as well as confer other beneficial effects on cognition, and potentially provides a simple way to modify the BDNF/trkB/histone acetylation pathways. Again, exercise appears to be a good way to achieve a healthspan enhancing effect.

#14. The DNA methylome as a biomarker for epigenetic instability and human aging.

One of the biggest problems with aging research is that reliable biomarkers for aging are controversial and limited in scope.

For example, telomere length which may not even be a good biomarker is not supposed to change in post-mitotic (non-dividing) cells. There is a great need for tissue-specific and cell-type specific aging biomarkers, so that the rates of aging may be determined at a greater resolution than the whole organism. Without good aging biomarkers it is very difficult to determine whether any specific potential therapy, drug, or dietary supplement really works.

Methylation of DNA is intimately involved in control of mammalian/vertebrate gene expression as part of a complex epigenetic regulatory system. We hypothesize that DNA methylation at cytosine–phosphate–guanine sites (CpGs), the "DNA methylome," evolved to increase stability of the differentiated state in somatic vertebrate cells, especially post-mitotic cells, which may have helped to increase longevity. Therefore, the DNA methylome may play a key role in human aging and be an ideal source of biomarkers aging. A new model that links the methylome to chronological age has been reported by Hannum *et al.* that accurately predicts age and rate of aging from the DNA methylation state of 71 markers in human blood samples. This model may make possible the development of new anti-aging therapeutics as well as more accurately assess the impact of anti-aging regimens, such as caloric restriction and drugs such as rapamycin. Furthermore, the model reveals information loss with increased age consistent with noise/unstable differentiation-based models of aging. The model may eventually lead to experiments to differentiate the contributions of biomolecular damage and noise/incomplete structural replication during aging.

Anyone pursuing anti-aging strategies would be advised to keep abreast of developments regarding new aging biomarkers in general and the DNA methylome biomarkers described in **Chapter 14** in particular. It is unlikely that one can develop potentially useful anti-aging drugs, supplements or other treatments without monitoring specific tissue rates of aging

#15. Rejuvenation of adult stem cells: is age-associated dysfunction epigenetic?

The dysfunctional changes of aging are generally believed to be irreversible due to the accumulation of molecular and cellular damage within an organism's somatic cells and tissues. However, the importance of potentially reversible cell signaling and epigenetic changes in causing dysfunction has not been thoroughly investigated.

Striking evidence that increased oxidative stress associated with hematopoietic stem cells (HSCs) from aging mice causes dysfunction has been reported. Forced expression of SIRT3, which activates the reactive oxygen species (ROS) scavenger superoxide dismutase 2 (SOD2) by de-acetylation to reduce oxidative stress, functionally rejuvenates mouse HSCs. These data, combined with numerous other reports, suggest that ROS act as a signal transducer to play a critical regulatory role in HSCs and at least in some other stem cells. It is likely that ectopic expression of SIRT3 restores homeostasis in gene expression networks sensitive to oxidative stress. This result was surprising because age-associated damage from impaired DNA repair had been thought to be irreversible in old HSCs.

The effect of up-regulated SIRT3 in HSCs is one of first examples in which intrinsic cellular aging, not apparently associated with changes in the micro-environment, was reversed. However, the stability of rejuvenation in the absence of continued supplemental SIRT3 expression was not investigated. These data are consistent with a hypothesis that potentially reversible processes, such as aberrant signaling and epigenetic drift, are relevant to cellular aging. If true, rejuvenation of at least some aged cells may be simpler than generally appreciated.

In these studies, the strong antioxidant N-acetylcysteine

restored HSC function in mice. This is the same N-acetylcysteine that blocks the beneficial effects of exercise in **Chapter 10**. Whether the continuous presence of SIRT3 or a strong antioxidant is needed for rejuvenation should be determined. Whether human HSCs behave the same way as murine HSCs needs to be studied as well, before anyone should consider using N-acetylcysteine to restore HSC function.

#16. Rejuvenation of aging hearts.

Specific subtle changes in regulation or activity of factors that maintain homeostasis and cell differentiation may play significant roles in mammalian aging. Drift resulting from reaching the end of an organism's developmental program might involve a specific ordered set of changes. Several studies have suggested that dysfunctional changes associated with aging in skeletal muscle, neurons, and hematopoietic stem cells may be caused by specific changes either in the extracellular environment or in intracellular regulatory networks and that such dysfunction may be reversible. On the basis these data, Loffredo *et al.* hypothesized that extrinsic circulating factors in young mice might reverse cardiac aging. Parabiosis, the surgical linking of circulations between old and young mice, was employed to identify an anti-hypertrophic factor (growth differentiation factor 11 [GDF-11]) that appears to rejuvenate aging murine hearts, raising exciting prospects for the development of anti-aging therapeutics. However, much work remains to be done to evaluate the utility of GDF-11 as a therapeutic rejuvenation factor. Similar rejuvenating factors for diverse tissues may exist as well and will hopefully be identified in the near future. This work is in its infancy, though a specific agonist of GDF-11 might be very useful to identify.

This work reinforces the possibility of restoring function to aged cells by specific manipulation of growth, trophic or other factors argues that many of the effects of aging itself may be due to the sequential breakdown of metastable biological networks. Identification of critical tissue specific rejuvenating factors may eventually extend healthspan and lifespan beyond their current limits.

Current perspective

Presently there are only a few clear-cut strategies to promote healthspan: moderate exercise, 7-8 hours of quality sleep, a smart balanced diet that avoids regular consumption of red meat (carnitine) and excessive sugar, and possibly dietary restriction. Although not covered in depth, extended periods of sitting should be avoided, since it has been correlated with as much as a 50% increase in all cause mortality, even in individuals who engage in leisure time physical activity.[1] Although not mentioned in the preceding chapters, attempting countermeasures for age-associated diseases that one is likely to experience is a reasonable idea. How to go about it? Reconciling known family medical history with personal genomics data (from such services as 23andme[2]) to establish some quantitative risk for critical age-associated diseases like for example type II diabetes, hypertension and Alzheimer's disease. Armed with foreknowledge of possible future diseases, it might be possible to use understanding of the underlying biological processes to select drugs and compounds found in supplements to delay or even forestall occurrence of pathological conditions. We already mentioned how some female carriers of BRCA mutations elect prophylactic surgery, but there are many possible examples. For

---

1 . Numerous studies support this conclusion. Reviewed by Dunstan DW, Howard B, Healy GN, Owen N. Too much sitting--a health hazard. Diabetes Res Clin Pract. 2012 97:368-76.

2 23andme has recently come under attack by the US FDA for providing personal medical data without proper medical guidance. Although personal genomic services should be very careful in providing interpretations of the data, a successful resolution of the current situation which reestablishes inexpensive access to one's own genetic data will benefit the public.

illustration, a person with a family history of type II diabetes, at high risk according to a personal genomics analysis or already exhibiting borderline elevated blood sugar levels (so-called prediabetes), might reasonably consult a physician to obtain a prescription for a low dose of metformin to be taken prophylactically. Metformin may have other benefits as well, as several of the chapters in this book will attest.

Please read our **Appendix: Applied Healthspan Engineering** for further discussion of some of the concepts described here, as well as for a 2010 perspective on the future of enhancing healthspan.

The Future

Some might apply the old adage about the last refuge of scoundrels to the art of predicting the future. But it is so easy and entertaining that more and more scientists can't help but succumb to temptation. And we are no better than our starry-eyed colleagues.

The biological sciences are in the midst of a revolution as biotechnology has advanced to the point, where unparallelled power to understand and alter biological material is within reach. These advances have brought into existence numerous new fields, the most important of which with regards to aging are systems biology, synthetic biology and regenerative biology.

Systems biology involves elucidating how the various parts of biological systems are connected and interact to make a living organism. In our appendix, we discuss the hierarchical nature of biological systems, which is just one of the many subjects of systems biology. Large scale initiatives to map the and model the genome, proteome, transcriptome, etc fall under its rubric. Especially important is the increasing ability to computationally model dynamic living systems on computers. <u>Employing the boatloads of new bioinformation to create increasingly realistic dynamic simulations of biological systems by leveraging large networks of powerful computers will be the key to progress.</u> Knowledge from this field is essential to be able to determine what changes might extend health and lifespan.

Synthetic biology is really a mishmash of biological fields that could just as well be called "Biotechnology 2.0." Essentially it encompasses every biotechnological advance beyond expressing or engineering a protein or drug. It involves using the knowledge obtained from systems biology to alter, program and create new organisms or biologically related entities. Hard core synthetic biology is based on using good engineering practices such as "standardized parts" to build interoperable systems, although the field is currently far broader in scope. Already it has been possible to program bacteria to act as photographic emulsion, or to identify and destroy some cancer cells (at least in a research lab). The scourge of cancer will likely fall, when synthetic biologists program a subset of one's own cells to seek out tumors by interacting with and querying cells for altered gene expression patterns, and then destroying the cancer. Synthetic biology will also provide the tools to program cells to perform tasks necessary to replace aging tissue most efficiently by encouraging regenerative processes.

Scientists studying regenerative biology and medicine seek to understand and then use knowledge of natural regenerative processes to replace damaged, missing or aged tissues. Regenerative biology is the most likely path leading to engineering greater lifespan and healthspan. Why? Because evolution itself has used regenerative processes to endow some organisms with a form of "negligible senescence," better known as immortality. Specifically, some organisms use their incredible regenerative powers to rebuild their body at will. Asexual planaria and hydra are the best understood examples. Hydra are capable not only of regenerating their bodies continuously, as old cells are simply sloughed off and replaced by new ones, but can actually reassembly themselves into an intact organism from their component cells after dissociation into single cells. Asexual planaria can rebuild their entire body from as small a fraction as 1/297. They carry pluripotent stem cells called neoblasts within their body, that can be activated to make any needed tissue or body structure at will. Research in ongoing to understand how they do this, perhaps so that our own cells can be similarly engineered using systems and synthetic biology.

It's not difficult to see that a confluence of genetics, regenerative biology, systems biology and synthetic biology will revolutionize medicine in the coming years. But like most revolutions, it will take time.

What is the current state of the art in regenerative biology? We know the exact molecular recipes to transform somatic cells into pluripotent stem cells, which can become any cell type. We are learning the recipes to change these pluripotent cells back into specific cell types, tissues and organs. Advances in the study of regeneration have already begun to identify key regulators of stem cell maintenance and regeneration itself. For example, mice lacking cell cycle inhibitor and tumor suppressor gene p21 have regenerative capacities not normally seen in mammals, such as healing a puncture wound to the earlobe using a structure called a blastema, where epithelial cells surround mesenchymal cells. The blastema is the hallmark of limb regeneration in newts and axolotls, and of planarian regeneration as well. A more substantial model of the cell and molecular networks involved in regeneration and in normal development will be of great utility to creating conventional therapeutics and then advanced therapies based on synthetic biology. Researchers around the globe are pursuing work to make regenerative medicine therapies a reality. One of us (ARM) founded Regenerative Sciences Institute, for the express purpose of catalyzing research combining systems, synthetic and regenerative biology and we encourage our readers to visit our website to learn more: www.regensci.org.

So why can't we have these wondrous advances now (which will be transformative outside of the narrow focus of aging)? 1) We are still gathering critical information and learning how to create dynamic models to integrate the information. 2) For therapies based on synthetic biology, we need gene therapy. Although molecular machines may be built such that they don't use the transcription/translation machinery of the host (DNA automatons have been developed to deliver a drug payload to cells carrying a specific receptor), and other informational biomolecules such as RNA may be used, DNA is the most natural choice as "wetware" for synthetic biology "programs." DNA-based programs invariably require some form of gene therapy, in which foreign DNA enters and interacts with human cells to accomplish a therapeutic purpose. Gene therapy, although technically possible[3] since the early 1980s, has been moving forward at a snail's pace, especially since the unexpected death of Jesse Gelsinger in a clinical trial in 1999. Although research has continued and numerous clinical trials are underway, only one gene therapy therapeutic, Glybera, which treats ipoprotein lipase deficiency, has been approved for human use, and that only in Europe. 3) Regulatory infrastructure to permit development of synthetic biology based, or even stem cell based therapies has not been established, and most importantly 4) Inertia from both regulators and from organizations (big Pharma) that have the huge financial resources to conduct clinical trials necessary to prove safety and efficacy.

So what will happen in the near term, that is, the "adjacent possible?' Better targeted small molecules, antibodies, proteins, formulated siRNA and perhaps more complex hybrid "smart"

---

3. This should be qualified by noting that technically possible does not imply practicality or safety.

molecules will be developed to treat numerous specific diseases, including those associated with aging. Some of these will be capable of enhancing regeneration as well as slowing some aspects of aging. Significant progress in extending health/life span will require the fruits of the revolution.

Section I

Repairing molecular damage

Chapter 1

Protein homeostasis as a clinical target for increased longevity?

Therapeutic intervention to extend health span and life span requires identification of critical mechanisms and specific targets. Loss of protein homeostasis (also known as proteostasis) has been implicated as a potential mechanism for aging.

Loss of Protein Homeostasis Has Been Hypothesized to Play a Critical Role in Aging

Accumulation of damaged proteins is considered to be a major mechanism by which homeostasis declines during aging. Protein homeostasis is defined as the set of transcriptional, translational, and posttranslational processes, including folding, trafficking, localization, and degradation, that maintain the correct amount, distribution, and structures of proteins within a cell. Mounting evidence suggests that maladaptive alterations in proteostasis, especially alterations in mechanisms that maintain protein quality control, such as chaperone function, unfolded protein response (UPR), ubiquitin/proteasome proteolysis, lysosomal function, macroautophagy, and chaperone-mediated autophagy (in mammals) contribute to aging (**Figure 1.1**)(1). Moreover, dysfunctional proteostasis plays a significant role in numerous degenerative pathologies, including Alzheimer's, Huntington's, and Parkinson's diseases.

One of the key targets of protein quality control is elimination of misfolded or modified proteins that oligomerize or aggregate. For example, in the nematode worm *Caenorhabditis elegans*, there is a pool of metastable proteins that begin to accumulate in misfolded conformations during early adulthood and lead to increasing

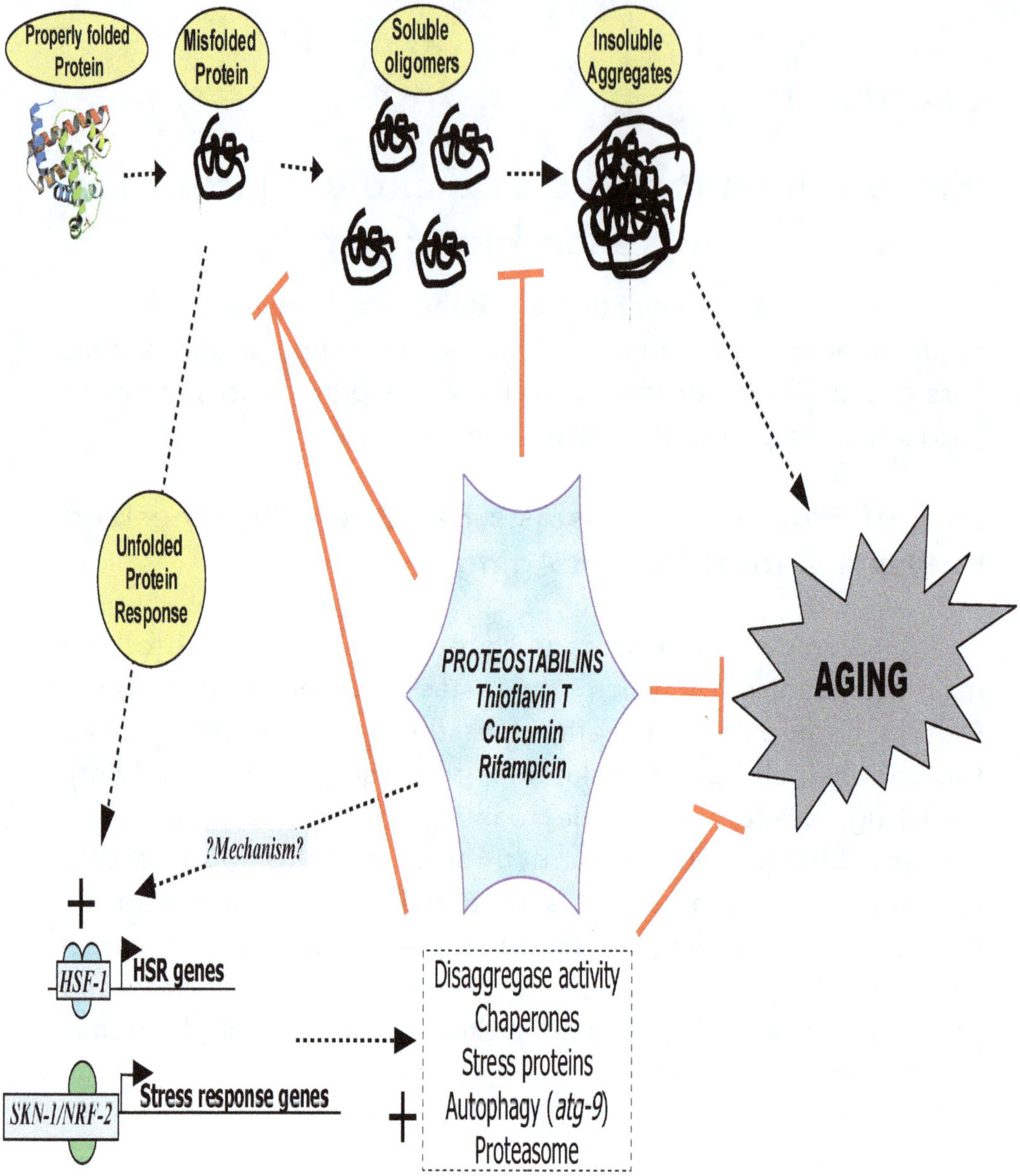

Figure. 1.1 **Pathways of proteostasis.** Protein misfolding and aggregation contribute to cellular dysfunction and loss of homeostasis. Alavez et al.(5) have demonstrated that thioflavin T and other small molecules ("proteostabilins") stabilize misfolded proteins to reduce aggregation and activate various protein quality control pathways. HSR, Heat shock response.

loss of protein quality control with age (2). Furthermore, several hundred mostly beta-sheet–enriched proteins that become more insoluble with age have been identified in adult *C. elegans* (3). Homologs of these proteins are over-represented in aggregates from humans with neurodegenerative diseases (3), suggesting that these proteins represent an important set of age-sensitive molecules. Their aggregation state may be a useful biomarker of compromised proteostasis associated with aging. Studies in rats demonstrating reduced chaperone levels with age in liver endoplasmic reticulum are consistent with the reduced protein quality control observed in worms, although more work is needed to understand the role of proteostasis in mammalian aging (4).

Alavez and Colleagues Report a Fascinating Study Demonstrating Extension of the Life Span of *C. elegans* by Protein-Stabilizing Compounds That Inhibit Protein Aggregation

To test the hypothesis that inhibition of protein misfolding and aggregation may help ameliorate dysfunctional proteostasis associated with aging, Alavez *et al.*(5) treated adult *C. elegans* with the flavanoid thioflavin T (ThT), a compound known to bind and inhibit aggregation of amyloids, such as those involved in Alzheimer's disease. Treatment with 50 or 100 mM ThT throughout adult life extended median life span by 60% and maximum life span by 43–78%. In addition, improved health was observed throughout adulthood, including reduced loss of movement associated with aging. Protein aggregate inhibitors curcumin and rifampicin also increased life span, but to a lesser extent (4,5). Combination of curcumin and ThT did not confer additional life span increase, suggesting that they act through a common mechanism, presumably stabilization and/or inhibition of protein aggregation.

To test directly whether ThT could suppress toxic effects of protein aggregation, worms were engineered to express temperature-sensitive aggregating proteins human beta-amyloid (amino acids 3-42) or polyglutamine (polyQ) in muscle. At 25C, the worms become paralyzed, due to the toxic effects of the aggregates. ThT suppressed paralysis, most likely by direct interaction with and stabilization of the pathological proteins. To show that this effect is general, ThT was used to suppress multiple worm temperature-sensitive mutations: gas-1 of mitochondrial complex 1, levamisole resistance in the alpha-subunit of the nicotinic receptor, unc-52 (perlecan), and unc-54 (myosin II heavy chain).

Dietary restriction (DR) is well known to increase life span in numerous species. ThT did not cause DR, as evidenced by a lack of a significant decrease in the pharyngeal pumping rate. However, the 100 mM dose of ThT was toxic to dietary- restricted worms, indicating that the combined effects of these treatments may be maladaptive in a dose-dependent manner. This effect suggests that a cautionary note should be sounded for humans attempting to combine antiaging regimens and compounds without data on their potential inter- actions. A lower dose (10 mM) of ThT extended the life span of DR-fed worms by an additional 24%, indicating that at least some of the mechanisms by which ThT extends life span are independent of DR.

To test involvement of protein quality control pathways, RNA interference (RNAi) knockdown and/or mutation of key proteostasis genes were studied. Thus, the chaperone Hsp16.(4,2), heat shock/protein quality control regulator HSF-1, and cell stress regulator SKN-1 (Nrf2-like) (**Figure 1.1**) are required for ThT to extend life span and to inhibit amyloid-mediated paralysis. These results indicate that protein quality control pathways also play a role in the ability of ThT to extend life span. Consistent with these results, ThT also increased expression of chaperones Hsp-6, Hsp16 (2), and Hsp-70. However, the relative importance of protein quality control pathways versus a direct effect on protein stability is yet to be determined.

The Big Picture for Aging: Are These Results Significant for Humans?

The results of the Alavez *et al.* paper lead to several important questions.

How much of aging can be attributed to dysfunction of proteostasis?

The short answer is that we do not know. The strongest consensus at this time would probably be that dysfunction in proteostasis is one of several mechanisms by which organisms age. Furthermore, this mechanism may play a greater role in postmitotic rather than in dividing cells. Quiescent cells do not undergo cell proliferation, which can dilute out misfolded or aggregated proteins.

Can up-regulation of protein quality control pathways prevent aging in worms?

Yes. Flies and worms that overexpress Hsp-70 and small heat shock proteins (sHSPs), such as Hsp16 and Hsp22, (6–11) have extended life spans. Furthermore, mutations that de- crease insulin signaling pathways or act as dietary restriction mimetics are associated with increased expression of Hsp-16 (12).

Does proteostasis play a similar role in humans or other mammals?

Organisms that are mostly post-mitotic as adults, such as *C. elegans* or *Drosophila melanogaster*, might be expected to be especially sensitive to alterations in protein quality control pathways. Consistent with this hypothesis is the observation that to date there are no reports of increased life span in transgenic mice that are engineered to express higher levels of proteostasis regulators or effectors. However, rapamycin, an inhibitor of the target of rapamycin (TOR) pathway, has been shown to extend life span in mice (13). Rapamycin increases autophagy, so there remains the possibility that autophagy plays a critical role in the rapamycin-mediated increase in longevity. On the other hand, an enhanced heat shock response may act to increase tumorigenesis and thereby shorten life span by buffering misfolded or mutant proteins found in cancer cells.

There are numerous reports of loss of proteostasis with increased age in mammals. In old rodents, endoplasmic reticulum (ER) chaperones, calnexin, protein disulfide isomerase (PDI), and binding immunoglobulin protein (BiP) have been observed to decrease with age in some organs and tissues (14–18) Furthermore, the UPR has been reported to be aberrant in aged animals (17,18) On the other hand, preliminary attempts to correlate 20S/26S proteasome activity with life span among endotherm vertebrate species has been unsuccessful (19), although a full exploration of comparative vertebrate protein quality control should be carried out.

More work is clearly needed to determine whether enhanced protein quality control can extend human health span or life span.

Medical Implications of This Work

The compounds investigated by Alavez *et al.* appear to extend longevity in *C. elegans* through two basic mechanisms:

(1) Biophysical stabilization of misfolded proteins and inhibition protein aggregation, and

(2) activation of protein quality control pathways though the heat shock response (HSR) regulator HSF-1 and cell stress regulator SKN1 (Nrf2- like).

We define compounds acting via both mechanisms as

"proteostabilins." As discussed above, the capacity of proteostabilins to increase life span or health span for humans remains to be established. However, approaches to stabilize protein con- formation are expected to benefit diseases caused by:

(1) Gain-of-function changes resulting in toxic protein aggregates, exemplified by Alzheimer's and Huntington's diseases, or

(2) loss-of-function mutations resulting in misfolded proteins such as Gaucher disease and cystic fibrosis (CF).

Compounds can be classified according to their mechanism of action, including pharmacological (chemical) chaperoning to maintain proper folding by direct interaction, kinetic stabilizers to shift equilibrium of a multimeric protein toward the most stable nonpathogenic form, and regulators of proteostasis that alter HPR, UPR, or lysosomal/autophagic pathways (20). A number of agents with these properties have been under active investigation for a number of diseases (**Table 1.1**).

Currently Available Drugs That Maintain Protein Stability, Prevent Protein Aggregation, or Stimulate Protein Stress Response Pathways

Extending life span by agents that help maintain proteostasis or act as proteostabilins remains hypothetical for mammals, including humans. However, there is evidence that such compounds may be useful to treat or prevent proteostatic diseases, many of which are associated with aging. Although no proteostatic therapeutic is Food and Drug Administration (FDA)–approved, there are a multitude of compounds being investigated for possible therapies of a variety of diseases where protein stability plays a critical role.

Direct protein stabilization

Agents that work through protein stabilization in the ER include 4-phenylbutyrate (4-PBA) and taurine-conjugated ursodeoxycholic acid (Table 1.1), which have been shown to increase glucose tolerance, normalize hyperglycemia, increase insulin sensitivity, and resolve fatty liver disease in a mouse model of type II

TABLE 1.1 Effects of Proteostabilins

Compound	Mechanism(s)	Condition
Thioflavin T (ThT)[5]	Protein stabilization, increased HSR (Hsf-1), stress response (Nrf2)	Life span (C. elegans)
Curcumin[5,27-31]	Protein stabilization, disaggregation, increased HSR (Hsf-1), stress-response (Nrf2)	Life span (C. elegans), Alzheimer disease, Parkinson disease
Rifampicin[5]	Protein stabilization, disaggregation, increased HSR (Hsf-1), stress-response (Nrf2)	Life span (C. elegans)
Epigallocatechin gallate (EGCG)[35-38]	Protein stabilization, disaggregation	Alzheimer disease, Parkinson disease
4-Phenylbutyrate[21,22]	ER protein stabilization	Diabetes type II, congenital nephrotic syndrome, CTFR
Diltiazem[25]	Ca^{2+} channel blocker increases HSR	Lysosomal storage disease
Taurine-conjugated ursodeoxycholic acid[21,22]	ER protein stabilization	Diabetes type II
Celastrol[24]	Increased Hsf1	Alzheimer disease
Carbamazipine[26]	Increased autophagy	α1-antitrypsin deficiency
Verapamil[25]	Ca^{2+} channel blocker increases HSR	Lysosomal storage disease
3,5-Dibromo-4-hydroxyphenyl–based stilbene/dihydrostilbene[23]	Protein kinetic stabilizer	Transthyretin-mediated amyloidoses
Coffee (extracts): polyphenols and caffeine[33,34]	Protein stabilization, disaggregation, increased stress response (Nrf2)	Alzheimer disease

HSR, Heat shock response; ER, endoplasmic reticulum; CTFR, cystic fibrosis transmembrane conductance regulator.

diabetes (leptin deficient, ob/ob) by relieving ER stress(21). 4-PBA also suppresses a destabilizing mutation in nephrin linked to congenital nephrotic syndrome in cultured human cells by restoring normal intracellular trafficking (22). 4-PBA partially restores the proteostasis of several misfolded soluble and transmembrane proteins, including the cystic fibrosis transmembrane conductance regulator (CFTR), for which there is an ongoing clinical trial.

Identification of compounds that act to kinetically stabilize transthyretin to prevent amyloidoses linked to malignant transhyretin aggregation is the subject of active research. Hundreds of such agents such as 3,5-dibromo-4-hydroxyphenyl–based stilbene and dihydrostilbene (**Table 1.1**) block aggregation of transthyretin in cell culture (23).

Modulation of proteostasis by modulation of HSR, UPR, and autophagy

Numerous compounds stimulate proteostasis by modulating protein quality control pathways. For example, celastrol (**Table 1.1**) increases HSF-1 levels in mammals to stimulate HSR and demonstrates protective activity in a murine model of Alzheimer's disease (24). Diltiazem and verapamil (**Table 1.1**), FDA-approved for hypertension, increase transcription of numerous cytoplasmic and ER chaperones by inhibiting L-type Ca++ channels in the plasma membrane. The resulting enhanced folding, trafficking, and activity of lysosomal enzymes helps suppress the pathologies associated with lysosomal storage diseases in cell culture models (25). Alpha-1-antitrypsin deficiency results from alpha-1-antitrypsin trapped in ER segments that are subject to autophagy. Carbamazepine (**Table 1.1**), an autophagy stimulator, has been shown to boost the capacity of the autophagic pathway to catabolize the aggregates of alpha-1-antitrypsin (26).

Proteostabilins

The compounds investigated by Alavez *et al.* appear to work by at least two mechanisms. As proteostabilins, they are expected to show similar activity in other systems. Curcumin has been shown to inhibit amyloid formation and toxicity (27) as well as be protective in a mouse model of Parkinson disease (28). A bioavailable nanoparticle preparation of curcumin had anti-amyloid activity in culture and a mouse model (29). Curcumin has also been shown to modestly extend life span in *C. elegans* (5), *D. melanogaster* (30) and mice (31). At least two other nutraceutical-based agents may act as proteostabilins. Polyphenols in coffee extracts suppress amyloid toxicity in a *C. elegans* model at least partially through the cell stress regulator SKN-1 (Nrf2)(32). Coffee itself has been shown to have anti-Alzheimer's disease activity in mice (33, 34). Epigallocatechin gallate (EGCG) (**Table 1.1**), a major component of green tea, has similar anti-amyloid/anti-Alzheimer's activities, including the ability to disaggregate amyloid fibrils and protect cultured cells (35,36). EGCG has potential therapeutic value for Parkinson disease as well (35). It has also been shown to increase life span significantly in *C. elegans*, upregulating SKN-1 (Nrf2) and Daf-16 (37,38) as well as modestly increasing life span in mice (31).

Conclusion

The mechanisms underlying proteostasis present attractive targets for pharmaceutical "proteostabilins" to extend health span and life span. Such efforts may be aided by further study of the specific roles played by proteostatic pathways in aging. Drugs to correct proteostatic dysfunction in diseases such as Alzheimer's disease are already under investigation or development. In the meantime, coffee or green tea may prove useful in staving off some neurodegenerative diseases.

References

1. Koga H, Kaushik S, Cuervo AM. Protein homeostasis and aging: The importance of exquisite quality control. Ageing Res Rev 2011;10:205–215.
2. Ben-Zvi A, Miller EA, Morimoto RI. Collapse of proteostasis represents an early molecular event in *Caenorhabditis elegans* aging. Proc Natl Acad Sci USA 2009;106:14914–14919.
3. David DC, Ollikainen N, Trinidad JC, Cary MP, Burlingame AL, Kenyon C. Widespread protein aggregation as an inherent part of aging in *C. elegans*. PLoS Biol 2010;8.
4. Erickson RR, Dunning LM, Holtzman JL. The effect of aging on the chaperone concentrations in the hepatic, endoplasmic reticulum of male rats: The possible role of protein misfolding due to the loss of chaperones in the decline in physiological function seen with age. J Gerontol A Biol Sci Med Sci 2006;61:435–443.
5. Alavez S, Vantipalli MC, Zucker DJS, Klang IM, Lithgow GJ. Amyloid-binding compounds maintain protein homeostasis during ageing and extend lifespan. Nature 2011;472:226–229.

6. Wang C, Li Q, Redden DT, Weindruch R, Allison DB. Statistical methods for testing effects on "maximum lifespan." Mech Ageing Dev 2004;125:629–632.
7. Walker GA, Lithgow GJ. Lifespan extension in *C. elegans* by a molecular chaperone dependent upon insulin-like signals. Aging Cell 2003;2:131–139.
8. Walker GA, White TM, McColl G, Jenkins NL, Babich S, Candido EP, Johnson TE, Lithgow GJ. Heat shock protein accumulation is upregulated in a long-lived mutant of *Caenorhabditis elegans*. J Gerontol A Biol Sci Med Sci 2001;56: B281–B287.
9. Morrow G, Samson M, Michaud S, Tanguay RM. Overexpression of the small mitochondrial Hsp22 extends drosophila life span and increases resistance to oxidative stress. FASEB J 2004;18:598–599.
10. Hsu A-L, Murphy CT, Kenyon C. Regulation of aging and age-related disease by daf-16 and heat-shock factor. Science 2003;300:1142–1145.
11. Morley JF, Morimoto RI. Regulation of longevity in *Caenorhabditis elegans* by heat shock factor and molecular chaperones. Mol Biol Cell 2004;15:657–664.
12. Wadhwa R, Takano S, Kaur K, Aida S, Yaguchi T, Kaul Z, Hirano T, Taira K, Kaul SC. Identification and characterization of molecular interactions between mortalin/mthsp70 and Hsp60. Biochem J 2005;391:185–190.
13. Harrison DE, Strong R, Sharp ZD, Nelson JF, Astle CM, Flurkey K, Nadon NL, Wilkinson JE, Frenkel K, Carter CS, Pahor M, Javors MA, Fernandez E, Miller RA. Rapamycin fed late in life extends lifespan in genetically heterogeneous mice. Nature 2009;460:392–395.
14. Hussain SG, Ramaiah KVA. Reduced eif-2 alpha phosphorylation and increased proapoptotic proteins in aging. Biochem Biophys Res Commun 2007;355:365–370.
15. Rabek JP, Boylston WH 3rd, Papaconstantinou J. Carbonylation of ER chaperone proteins in aged mouse liver. Bio- chem Biophys Res Commun 2003;305:566–572.
16. Nuss JE, Choksi KB, DeFord JH, Papaconstantinou J. Decreased enzyme activities of chaperones PDI and BiP in aged mouse livers. Biochem Biophys Res Commun 2008;365:355– 361.
17. Naidoo N, Ferber M, Master M, Zhu Y, Pack AI. Aging impairs the unfolded protein response to sleep deprivation and leads to pro-apoptotic signaling. J Neurosci 2008;28:6539–6548.
18. Paz Gavilan M, Vela J, Castano A, Ramos B, del Rio JC, Vitorica J, Ruano D. Cellular environment facilitates protein accumulation in aged rat hippocampus. Neurobiol Aging 2006;27:973–982.
19. Salway KD, Page MM, Faure PA, Burness G, Stuart JA. Enhanced protein repair and recycling are not correlated with longevity in 15 vertebrate endotherm species. Age (Dordr) 2011;33:33–47.
20. Balch WE, Morimoto RI, Dillin A, Kelly JW. Adapting proteostasis for disease intervention. Science 2008;319:916–919.

21. Ozcan U, Yilmaz E, Ozcan L, Furuhashi M, Vaillancourt E, Smith RO, Gorgun CZ, Hotamisligil GS. Chemical chaperones reduce er stress and restore glucose homeostasis in a mouse model of type 2 diabetes. Science 2006;313:1137–1140.

22. Liu XL, Done SC, Yan K, Kilpelainen P, Pikkarainen T, Tryggvason K. Defective trafficking of nephrin missense mutants rescued by a chemical chaperone. J Am Soc Nephrol 2004;15:1731–1738.

23. Connelly S, Choi S, Johnson SM, Kelly JW, Wilson IA. Structure-based design of kinetic stabilizers that ameliorate the transthyretin amyloidoses. Curr Opin Struct Biol 2010;20:54–62.

24. Paris D, Ganey NJ, Laporte V, Patel NS, Beaulieu-Abdelahad D, Bachmeier C, March A, Ait-Ghezala G, Mullan MJ. Reduction of beta-amyloid pathology by celastrol in a transgenic mouse model of Alzheimer's disease. J Neuroinflammation 2010;7:17–17.

25. Mu T-W, Fowler DM, Kelly JW. Partial restoration of mutant enzyme homeostasis in three distinct lysosomal storage disease cell lines by altering calcium homeostasis. PLoS Biol 2008;6.

26. Hidvegi T, Ewing M, Hale P, Dippold C, Beckett C, Kemp C, Maurice N, Mukherjee A, Goldbach C, Watkins S, Michalopoulos G, Perlmutter DH. An autophagy-enhancing drug promotes degradation of mutant a1-anti-trypsin and reduces hepatic fibrosis. Science 2010;329:229–232.

27. Mishra S, Mishra M, Seth P, Sharma SK. Tetrahydrocurcumin confers protection against amyloid b-induced toxicity. Neuroreport 2010.

28. Khuwaja G, Khan MM, Ishrat T, Ahmad A, Raza SS, Ashafaq M, Javed H, Khan MB, Khan A, Vaibhav K, Safhi MM, Islam F. Neuroprotective effects of curcumin on 6-hydroxydopamineinduced Parkinsonism in rats: Behavioral, neurochemical and immunohistochemical studies. Brain Res 2011;1368:254–263.

29. Ray B, Bisht S, Maitra A, Maitra A, Lahiri DK. Neuroprotective and neurorescue effects of a novel polymeric nano- particle formulation of curcumin (nanocurcTM) in the neuronal cell culture and animal model: Implications for Alzheimer's disease. J Alzheimers Dis 2011;23:61–77.

30. Lee K-S, Lee B-S, Semnani S, Avanesian A, Um C-Y, Jeon H- J, Seong K-M, Yu K, Min K-J, Jafari M. Curcumin extends life span, improves health span, and modulates the expression of age-associated aging genes in *Drosophila melanogaster*. Rejuvenation Res 2010;13:561–570.

31. Kitani K, Osawa T, Yokozawa T. The effects of tetra-hydrocurcumin and green tea polyphenol on the survival of male C57Bl/6 mice. Biogerontology 2007;8:567–573.

32. Dostal V, Roberts CM, Link CD. Genetic mechanisms of coffee extract protection in a Caenorhabditis elegans model of {beta}-amyloid peptide toxicity. Genetics 2010;186:857–866.

33. Eskelinen MH, Kivipelto M. Caffeine as a protective factor in dementia and

Alzheimer's disease. J. Alzheimers Dis 2010; 20 (Suppl 1):S167–S174.
34. Arendash GW, Mori T, Cao C, Mamcarz M, Runfeldt M, Dickson A, Rezai-Zadeh K, Tane J, Citron BA, Lin X, Echeverria V, Potter H. Caffeine reverses cognitive impairment and decreases brain amyloid-beta levels in aged Alzheimer's disease mice. J. Alzheimers Dis 2009;17:661–680.
35. Mandel SA, Amit T, Weinreb O, Reznichenko L, Youdim MBH. Simultaneous manipulation of multiple brain targets by green tea catechins: a potential neuroprotective strategy for Alzheimer's and Parkinson's diseases. CNS Neurosci Ther 2008;14:352–365.
36. Meng F, Abedini A, Plesner A, Verchere CB, Raleigh DP. The flavanol (+)-epigallocatechin 3 gallate inhibits amyloid formation by islet amyloid polypeptide, disaggregates amyloid fibrils, and protects cultured cells against iapp-induced toxicity. Biochemistry 2010;49:8127–8133.
37. Abbas S, Wink M. Epigallocatechin gallate from green tea (Camellia sinensis) increases lifespan and stress resistance in *Caenorhabditis elegans*. Planta Med 2009;75:216–221.
38. Zhang L, Jie G, Zhang J, Zhao B. Significant longevity-extending effects of EGCG on *Caenorhabditis elegans* under stress. Free Radic Biol Med 2009;46:414–421.

Section II

Repairing intracellular substructures (mitochondria, telomeres)

Chapter 2

Master switch of mitochondrial biogenesis: a clinical target for health span enhancement?

Therapeutic intervention to extend health span and life span requires identification of critical mechanisms and specific targets. Two mechanisms that have been implicated in aging are mitochondrial dysfunction and replicative senescence:

1. Damage to the mitochondria has been hypothesized to play a critical role in cellular and organismal aging. Mitochondrial theories of aging hypothesize that reactive oxygen species (ROS) generated during aerobic respiration damage proteins, lipids and DNA in the mitochondria (1). Accumulated damage to the respiratory chain increases the generation of ROS, which contributes to mitochondrial dysfunction either directly, through faulty mitochondrial DNA (mtDNA)- encoded proteins (2) or via the lysosomal–mitochondrial axis (3) in which damage to both organelles combines to reduce cell function. Mitochondrial homeostasis has been incorporated into these models, most often emphasizing how ROS-mediated damage attenuates autophagy or biogenesis, thus leading to cellular dysfunction.

2. Loss of telomere length in somatic cells presents a potential barrier to life span and health span enhancement. Most somatic mammalian cells lack telomerase activity, which results in a limited replicative life span called the "Hayflick limit," named after its discoverer Leonard Hayflick (4). Telomeres shorten with each cell division until the cell senses the short telomeres as DNA damage, stops dividing, and

enters a "replicative senescent" state. The absence of telomerase in somatic cells has been hypothesized to be an anticancer adaptation (5), because it must be overcome for a transformed cell to form a clinically relevant tumor. Although, the connection between short telomeres and normal human aging is yet to be established conclusively, the absence of telomerase in most somatic cells presents a theoretical barrier to strategies for engineered negligible senescence (SENS) that would have to be overcome at least transiently (6). The Hayflick limit and mitochondrial dysfunction have largely been seen to be unconnected until now.

Sahin and colleagues (7) report a fascinating study linking telomere and mitochondrial dysfunction. So-called "G4" transgenic mice accumulated short telomeres over four generations due to the absence of either the telomerase reverse transcriptase (Tert) or telomerase RNA component (Terc) genes. As expected, the mice accumulated pathologies associated with proliferating cells ("stem cell depletion"). However, they also exhibited cardiomyopathy and liver dysfunction (glucose intolerance and reduced capacity for detoxification), pathologies in organs with predominately non-proliferating cells. Hematopoietic stem cells (HSCs), liver cells, and heart cells from G4 mice had reduced mitochondrial function, including reduced adenosine triphosphate (ATP) synthesis, fewer mitochondria, reduced levels of some ROS detoxifying enzymes, and increased carbonylated proteins and ROS levels. Together, these findings indicate a profound level of mitochondrial dysfunction similar to that seen in old animals or humans.

Messenger RNA (mRNA) profiling revealed that master transcriptional regulators, peroxisome proliferator-activated receptor-gamma (PPAR-gamma), co-activator 1 a and b (PGC-1a and PGC-1b) were significantly reduced. These genes play a critical role in mitochondrial biogenesis (see ref. 12, below). G4 mice bearing reduced levels of PGC-1a/b exhibit impaired gluconeogenesis, cardiomyopathy, and reduced ability of HSCs to reconstitute bone marrow. Adenoviral expression of the PGC-1a gene in the liver of G4 mice restored mitochondrial function and gluconeogenesis. The authors hypothesize that activation of p53 due to the presence of dysfunctional telomeres is responsible for the repression of PGC-1a and PGC-1b. To support this contention, they found that p53 binds the PGC-1a and PGC-1b promoters and conversely that G4 mice lacking p53 exhibit substantial PGC-1a and PGC-1b expression. In these mice (p53-/- xTerc-/-), reduction of p53 partially rescues the mitochondrial, heart failure, and gluconeogenesis phenotypes that result from the short telomeres. Taken together these data strongly suggest that short telomeres, and perhaps other DNA damage or even other types of cell stress can cause p53-mediated suppression of PGC-1a and PGC-1b, which, in turn, leads to mitochondrial dysfunction similar to that seen in normal aging.

It would be interesting to test whether downregulation of PGC-1a or PGC-1b can partially explain the shorter life span seen in transgenic mice that express hyperactive p53 (8,9). Mitochondrial dysfunction resulting from repression of PGC-1a and PGC-1b is not unexpected because these master regulators of mitochondrial biogenesis are necessary for mitochondrial homeostasis and are known to be expressed at lower levels in old animals. These proteins are positioned at a signaling nexus that integrates cellular response to ROS and mitochondrial metabolism. Given the strong phenotypes seen in this paper, we hypothesize that differential regulation of mitochondrial biogenesis via PGC-1a and PGC-1b signaling may play a more primary role than previously thought in increasing mitochondrial dysfunction with age and represent a good target for health span enhancement. In turn, signaling from the mitochondria may play a more fundamental role in aging than previously assumed as well. For example, Durieux *et al.* (10) recently showed that electron transport chain (ETC)-mediated longevity in *Caenorhabditis elegans* is communicated to distal cells by a mitochondrial stress-triggered unfolded protein response (UPR) by an unidentified mitokine.

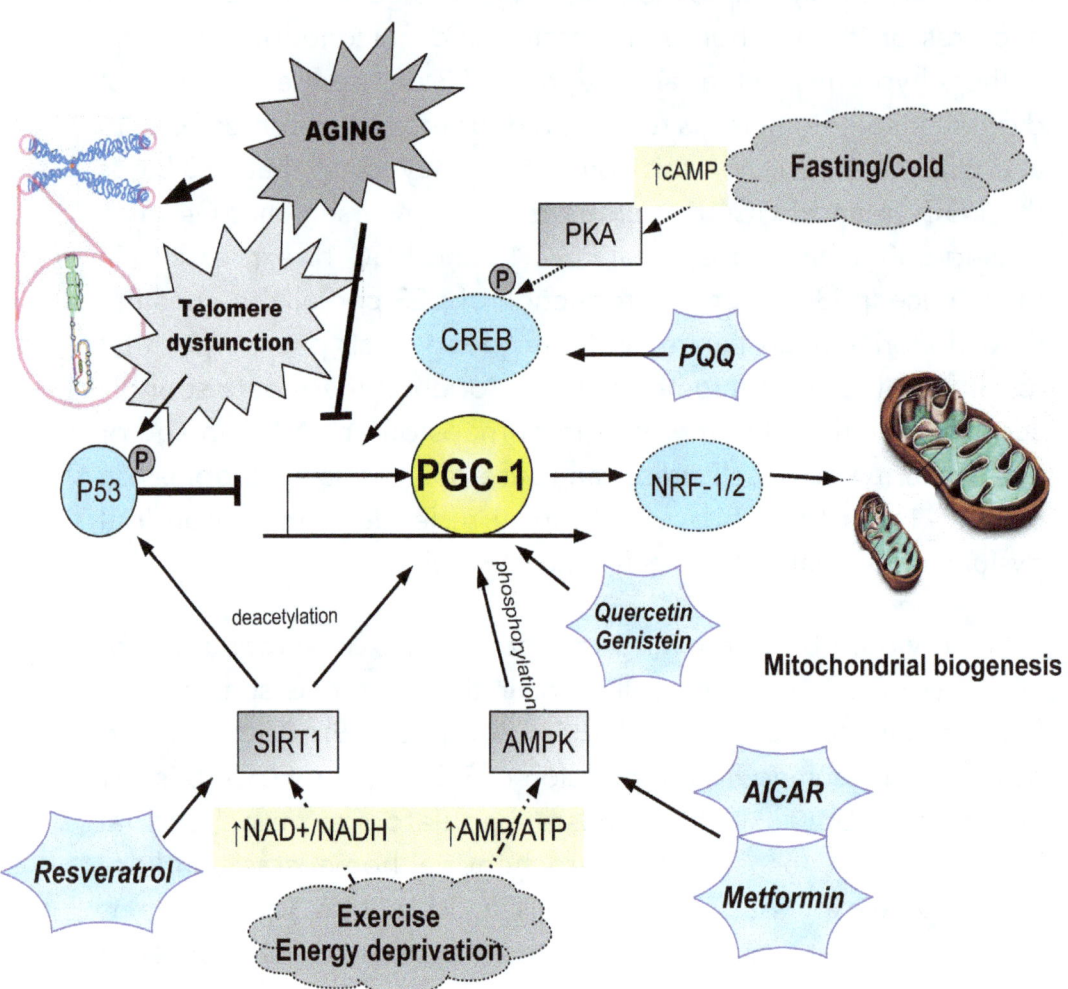

Figure 2.1 **Peroxisome proliferator-activated receptor-γ co-activator 1 (PGC-1) signaling pathways control mitochondrial biogenesis.** Cellular stresses associated with aging, including telomere dysfunction, dietary excess and inactivity down-regulate activity of master regulators PGC-1α/β through p53. Exercise, fasting, and exposure to cold upregulate PGC-1 through stimulation of sirtuin 1 (SIRT1), adenosine 5′-monophosphate-activated protein kinase AMPK, and cyclic AMP response element-binding (CREB). Active PGC-1 stimulates mitochondrial biogenesis through increased NRF1/2. Several promising pharmaceutical interventions shown to modulate PGC-1 are shown (hexagonal stars).

Medical Implications of This Work

The work of Sahin et al. (7) suggests that p53, telomerase, and/or PGC-1a and PGC-1b may be interesting targets for therapeutic intervention for aging-associated diseases. However, the association of p53 inactivation or telomerase reactivation with potential tumorigenesis should inspire great caution. Therefore, we focus below on PGC-1a and PGC-1b. That forced expression of moderate amounts of PGC-1a in skeletal muscle protects aging mice from sarcopenia, loss of bone mineral density, increased inflammation, and loss of insulin sensitivity (11) is further reason for exploring these targets.

Regulation of PGC-1a and PGC-1b

What do we know? Many reports support the role of PGC- 1a and PGC-1b as master regulators of mitochondrial biogenesis. Overexpression of PGC-1a or PGC-1b in cultured cells or in transgenic mice results in massive increases in mitochondrial content. Knockout of both PGC-1a or PGC-1b leads to death of newborn mice within a few days, and these mice have fewer mitochondria (12) Skeletal muscle is significantly affected by overexpression of these proteins: PGC-1a inhibits muscle atrophy in animals that do not exercise (13) and confers strongly enhanced exercise capability on such mice (14), and PGC-1b improves insulin resistance in skeletal muscle (15). Thus, it is likely that stimulation of PGC-1a and PGC-1b will benefit health span.

PGC-1a has a potent amino-terminal transcriptional activation domain, whereby it interacts with chromatin re- modeling co-activator complexes, including GCN5, SRC-1, and CBP/p300. It also binds the promoter regions of many transcription factors and nuclear hormone receptors including PPAR-g, ERRa, which controls fatty acid oxidation, and NRF-1, YY1, PPAR-a, and MEF2C, which act on the respiratory chain. PGC-1b lacks the RNA-processing domain of PGC-1a, but retains much of the same ability to stimulate multiple nuclear respiratory genes. PGC-1b promotes a much higher level of coupled respiration than PGC-1a because of differences in proton leakage (12).

PGC-1a and PGC-1b stimulate mitochondrial biogenesis by stimulating expression of these downstream targets.

PGC-1's are regulated by Sirtuin 1 and adenosine-5'-monophosphate–activated kinase PGC-1a is activated by the nicotinamide adenine dinucleotide (NAD)-dependent deacetylase sirtuin 1 (SIRT1). The SIRT1 homolog SIR2 is associated with increased replicative life span in yeast and increased longevity in *C. elegans* (16) SIRT1 deacetylates multiple lysine residues on PGC-1a, promoting mitochondrial fatty acid oxidation in response to low glucose. Adenosine-5'-monophosphate (AMP)-activated protein kinase (AMPK), a sensor that is activated upon energy depletion in muscle and is known to stimulate mitochondrial biogenesis as well, (17) phosphorylates PGC-1a on threonine-177 and serine-538. AMPK mediates at least some of its effects to stimulate mitochondrial biogenesis via PGC-1a. AMPK is downregulated during aging (18) SIRT and AMPK pathways cooperate to mediate calcium-dependent mitochondrial proliferation in myocytes. Therefore, stimulation of SIRT1 and AMPK should help activate PGC-1 proteins (**Figure 2.1**)

Current availability of drugs or activities that stimulate PGC-1a or PGC-1b

Given the potential benefits of augmented PGC-1a and PGC-1b function, new drugs to increase expression or activity of these proteins are expected to have clinical benefit. Several existant drugs or nutraceuticals known to increase the activity of AMPK or SIRT1 also stimulate PGC-1 proteins, at least indirectly. For example, metformin is known to stimulate AMPK via inhibition of AMP deaminase (19), and multiple studies indicate stimulation of PGC-1 (20,21). Resveratrol, which stimulates SIRT1, possibly indirectly (22), and indirectly stimulates AMPK, upregulates PGC-1a activity (23, 24). Dietary supplementation with pyrroloquinoline quinone (PQQ), a redox cofactor, increases mitochondrial biogenesis and function in mice (25, 26). In these mice, PGC-1a is elevated and required for PQQ effects. PGC-1a is stimulated through increased phosphorylation of cyclic (c) AMP response element-binding (CREB) on Ser-133, which itself is required for PGC-1a upregulation (26). Additional compounds that stimulate PGC-1a include genistein and other isoflavones (27) and quercetin (28). Finally, AICAR (5-aminoimidazole-4-carboxamide-1-b-D-ribofuranoside), an experimental drug that increases AMPK activity, also stimulates PGC-1a (24, 29).

Moderate exercise may be the safest means to increase PGC-1. Elevated PGC-1 protein expression has been reported in exercised animals and humans, with P38/MAPK and calcium calmodulin protein kinases playing key roles (30). Although high-intensity interval training raises both SIRT1 and PGC-1a levels (31), low-volume sprint interval exercise in cyclists appears to achieve similar benefits (32) Intriguingly, recent work suggests that the UPR plays a key role in skeletal muscle adaptation to exercise. This effect depends on PGC-1a cooperating with ATF6a to induce the UPR (33). Given the role of the UPR to mediate organismal stress signaling (10), PGC-1 activity may play an even more fundamental role in aging than currently appreciated.

Conclusion

The pioneering work of Sahin *et al.* (7) connecting replicative senescence with mitochondrial dysfunction, suggests that the master regulators of mitochondrial biogenesis PGC-1a and PGC-1b play an important role in aging-related pathologies. Increased expression of these genes undoubtedly explains the benefits of simple exercise. Conversely, reduction in their activity is contributing to the global pandemic of metabolic syndrome associated with sedentary lifestyles and over-abundant nutrition. Development of new drugs that specifically augment PGC-1 activity may prove beneficial to prevent and reverse some of the detrimental effects of mitochondrial dysfunction seen with aging.

References

1. Gruber J, Schaffer S, Halliwell B. The mitochondrial free radical theory of ageing—where do we stand? Front Biosci 2008;13:6554–6579.
2. de Grey ADNJ. A proposed refinement of the mitochondrial free radical theory of aging. Bioessays 1997;19:16–166.
3. Terman A, Kurz T, Navratil M, Arriaga EA, Brunk UT. Mitochondrial turnover and aging of long-lived postmitotic cells: The mitochondrial–lysosomal axis theory of aging. Antioxid Redox Signal 2010;12:503–535.
4. Cristofalo VJ, Lorenzini A, Allen R, Torres C, Tresini M. Replicative senescence: A critical review. Mech Ageing Dev 2004;125:827–848.
5. Rodier F, Campisi J. Four faces of cellular senescence. J Cell Biol 2011;192:547–556.
6. de Grey A, Rae M. Ending Aging: The Rejuvenation Breakthroughs That Could Reverse Human Aging in Our Lifetime. St. Martin's Press, New York, 2007.
7. Sahin E, Colla S, Liesa M, Moslehi J, Muller FL, Guo M, Cooper M, Kotton D, Fabian AJ, Walkey C, Maser RS, Tonon G, Foerster F, Xiong R, Wang YA, Shukla SA, Jaskelioff M, Martin ES, Heffernan TP, Protopopov A, Ivanova E, Mahoney JE, Kost-Alimova M, Perry SR, Bronson R, Liao R, Mulligan R, Shirihai OS, Chin L, DePinho RA. Telomere dysfunction induces metabolic and mitochondrial compromise. Nature 2011;470:359–365.
8. Maier B, Gluba W, Bernier B, Turner T, Mohammad K, Guise T, Sutherland A, Thorner M, Scrable H. Modulation of mammalian life span by the short isoform of p53. Genes Dev 2004;18:306–319.
9. Tyner SD, Venkatachalam S, Choi J, Jones S, Ghebranious N, Igelmann H, Lu X, Soron G, Cooper B, Brayton C, Hee Park S, Thompson T, Karsenty G, Bradley A, Donehower LA. P53 mutant mice that display early ageing-associated phenotypes. Nature 2002;415:45–53.
10. Durieux J, Wolff S, Dillin A. The cell-non-autonomous nature of electron transport chain mediated longevity. Cell 2011;144:79–91.
11. Wenz T, Rossi SG, Rotundo RL, Spiegelman BM, Moraes CT. Increased muscle pgc-1a expression protects from sarcopenia and metabolic disease during aging.

Proc Natl Acad Sci USA 2009;106:20405–20410.

12. Scarpulla RC. Metabolic control of mitochondrial biogenesis through the PGC-1 family regulatory network. Biochim Biophys Acta Mol Cell Res 2010 (in press) doi:10(1)016/ j.bbamcr 2010.09.019.

13. Brault JJ, Jespersen JG, Goldberg AL. Peroxisome proliferator-activated receptor gamma coactivator 1a or 1b overexpression inhibits muscle protein degradation, induction of ubiquitin ligases, and disuse atrophy. J Biol Chem 2010;285:19460–19471.

14. Calvo JA, Daniels TG, Wang X, Paul A, Lin J, Spiegelman BM, Stevenson SC, Rangwala SM. Muscle-specific expression of PPARg coactivator-1a improves exercise perfor- mance and increases peak oxygen uptake. J Appl Physiol 2008;104:1304–1312.

15. Wright LE, Brandon AE, Hoy AJ, Forsberg G, Lelliott CJ, Reznick J, Lofgren L, Oscarsson J, Stromstedt M, Cooney GJ, Turner N. Amelioration of lipid-induced insulin resistance in rat skeletal muscle by overexpression of pgc-1a involves reductions in long-chain acyl-coa levels and oxidative stress. Diabetologia 2011 (in press) doi:10(1)007/s00125-011-2068-x..

16. Donmez G, Guarente L. Aging and disease: Connections to sirtuins. Aging Cell 2010;9:285–290.

17. Jornayvaz FR, Shulman GI. Regulation of mitochondrial biogenesis. Essays Biochem 2010;47:69–84.

18. Reznick RM, Zong H, Li J, Morino K, Moore IK, Yu HJ, Liu Z, Dong J, Mustard KJ, Hawley SA, Befroy D, Pypaert M, Hardie DG, Young LH, Shulman GI. Aging-associated reductions in AMP-activated protein kinase activity and mitochondrial biogenesis. Cell Metab 2007;5:151–156.

19. Ouyang J, Parakhia RA, Ochs RS. Metformin activates AMP kinase through inhibition of AMP deaminase. J Biol Chem 2011;286:1–11.

20. Suwa M, Egashira T, Nakano H, Sasaki H, Kumagai S. Metformin increases the PGC-1alpha protein and oxidative enzyme activities possibly via AMPK phosphorylation in skeletal muscle *in vivo*. J Appl Physiol 2006;101:1685–1692.

21. Gundewar S, Calvert JW, Jha S, Toedt-Pingel I, Ji SY, Nunez D, Ramachandran A, Anaya-Cisneros M, Tian R, Lefer DJ. Activation of AMPK by metformin improves left ventricular function and survival in heart failure. Circ Res 2009;104: 403–411.

22. Beher D, Wu J, Cumine S, Kim KW, Lu S, Atangan L, Wang M. Resveratrol is not a direct activator of SIRT1 enzyme activity. Chem Biol Drug Des 2009;74:619–624.

23. Chaudhary N, Pfluger PT. Metabolic benefits from SIRT1 and SIRT1 activators. Curr Opin Clin Nutr Metab Care 2009;12:431–437.

24. Yu L, Yang SJ. Amp-activated protein kinase mediates activity-dependent regulation of peroxisome proliferator-activated receptor gamma coactivator-1alpha and nuclear respiratory factor 1 expression in rat visual cortical neurons.

Neuroscience 2010;169:23–38.

25. Stites T, Storms D, Bauerly K, Mah J, Harris C, Fascetti A, Rogers Q, Tchaparian E, Satre M, Rucker RB. Pyrroloqui- noline quinone modulates mitochondrial quantity and function in mice. J Nutr 2006;136:390–396.

26. Chowanadisai W, Bauerly KA, Tchaparian E, Wong A, Cortopassi GA, Rucker RB. Pyrroloquinoline quinone stimulates mitochondrial biogenesis through camp response element-binding protein phosphorylation and increased PGC-1a expression. J Biol Chem 2010;285:142–152.

27. Rasbach KA, Schnellmann RG. Isoflavones promote mitochondrial biogenesis. J Pharmacol Exp Therapeutics 2008;325: 536–543.

28. Davis JM, Murphy EA, Carmichael MD, Davis B. Quercetin increases brain and muscle mitochondrial biogenesis and exercise tolerance. Am J Physiol Regul Integr Comp Physiol 2009;296:R1071–R1077.

29. Tadaishi M, Miura S, Kai Y, Kawasaki E, Koshinaka K, Kawanaka K, Nagata J, Oishi Y, Ezaki O. Effect of exercise intensity and aicar on isoform-specific expressions of murine skeletal muscle PGC-1a mRNA: a role of b2-adrenergic receptor activation. Am J Physiol Endocrinol Metab 2011;300: E341–E349.

30. Lira VA, Benton CR, Yan Z, Bonen A. Pgc-1a regulation by exercise training and its influences on muscle function and insulin sensitivity. Am J Physiol Endocrinol Metab 2010;299: E145–E161.

31. Gurd BJ, Perry CGR, Heigenhauser GJF, Spriet LL, Bonen A. High-intensity interval training increases SIRT1 activity in human skeletal muscle. Appl Physiol Nutr Metab 2010;35: 350–357.

32. Psilander N, Niklas P, Wang L, Li W, Westergren J, Jens W, Tonkonogi M, Michail T, Sahlin K, Kent S. Mitochondrial gene expression in elite cyclists: Effects of high-intensity interval exercise. Eur J Appl Physiol 2010;110:597–606.

33. Wu J, Ruas JL, Estall JL, Rasbach KA, Choi JH, Ye L, Bostrom P, Tyra HM, Crawford RW, Campbell KP, Rutkowski DT, Kaufman RJ, Spiegelman BM. The unfolded protein response mediates adaptation to exercise in skeletal muscle through a PGC-1a/ATF6a complex. Cell Metab 2011;13:160–169.

Chapter 3

Ectopic expression of telomerase safely increases health span and life span

The absence of telomerase from somatic cells of mammals has significant consequences for aging. First, it limits the number of potential cell divisions and in so doing sets limits on both life span and cancer cell proliferation. Second, shortened telomeres are known to result in physiological dysfunction, including playing a role in human diseases such as Werner syndrome and ataxia telangiectasia. Ectopic expression of the catalytic subunit of telomerase, telomerase reverse transcriptase (TERT), has been reported to extend life span by as much as 40% in cancer-resistant mice. On the other hand, ectopic expression of TERT promotes cancer in normal mice. However, transient induction of TERT by an astragalus-derived compound increases health span without an apparent increase in cancer incidence. Ectopic expression of TERT using adeno-associated virus serotype 9 (AAV9)-based gene therapy in adult mice increases both health span and life span without increasing cancer incidence. Available evidence suggests that increases in life span may require both elongated telomeres and the continuous presence of telomerase to stimulate the WNT/beta-catenin signaling pathway. The recent observation that WNT/beta-catenin signaling can stimulate TERT expression raises the possibility of a positive feedback loop between TERT and WNT/beta-catenin. Such a positive feedback loop implies that safety must be carefully considered in the development of drugs that stimulate telomerase activity.

Introduction

Epigenetic changes play a critical role in effecting the age-associated decline of function that leads to mortality in animals (1). Whether these changes ultimately result from accumulation of molecular damage or from phenotypic instability associated with incomplete maintenance of cellular state is under investigation. An example of the latter possibility is the shortening of telomeres with each cell division in somatic cells of many animals due to the absence of the telomere-regenerating enzyme telomerase. The absence of telomerase in somatic cells limits the number of potential cell divisions (replicative or Hayflick limit) and has been hypothesized to be a mechanism that protects organisms from tumorigenesis. Cells that reach their replicative limit undergo senescence, which involves activation of an altered phenotypic program that may promote organismal dysfunction. Cancers overcome this potential blockade by either expressing telomerase (about 90% of the time) or activating the alternative lengthening of telomeres (ALT) recombination-based pathway to become immortal.

The absence or low-level expression of telomerase in somatic cells poses an absolute roadblock to Strategies for Engineered Negligible Senescence (SENS). For example, recently it was shown that an asexual planarian species that reproduces by a regenerative fission-like process constitutively expresses telomerase in its adult stem cells (neoblasts), while an almost genetically identical sexually reproducing planarian expresses telomerase only in the germ cells. The adult stem cell telomerase activity helps confer a form of immortality (SENS by evolution) on the asexual planarian strain (2).

The absence of telomerase/telomere shortening in somatic cells plays a controversial role in mammalian aging. On the one hand, genetic knockout of telomerase function in mice has little noticeable effect on the aging of first-generation mutants. Serious phenotypic consequences are seen only in the fourth through sixth generations of such mutants when premature aging-associated phenotypes appear. This is be- cause the normal length of mouse telomeres is sufficient for several mouse life spans, including all of the cell divisions associated with development. On the other hand, ectopic expression of the catalytic subunit of telomerase (telomerase reverse transcriptase, TERT) in epithelial cells has been re- ported to extend life span by up to 40% in mice engineered to be cancer-resistant (3). Unfortunately, ectopic expression of TERT in wild-type mice or mutations in human TERT increase cancer risk (4–8). There is evidence that active telomerase/long telomeres protect cells from the metabolic and mitochondrial compromise that occurs when shortened telomeres induce p53, which in turn represses the promoters of master metabolic/mitochondrial regulators peroxisome proliferator-activated receptor-c coactivator (PGC)-1a and PGC-1b (9). Ironically, shortened telomeres also result in increased cancer rates, probably due to increased genomic instability (10,11). Consistent with a homeostatic mechanism is the observation that telomerase reactivation has been shown to partially reverse tissue degeneration in aged telomerase- deficient mice (fourth generation) (12).

There is a paradox here: Mouse chromosomes possess enough reserve telomere length to fuel cell divisions for up to six organismal generations, yet mice apparently have at least a subset of cells in which dysfunction is linked to shorter telomeres and/or the absence of telomerase within a single life span. This paradox relates to the critical question of whether sufficient clinical benefit could result from ectopic telomerase expression in human aging and in diseases associated with shortened telomeres such as autoimmune diseases (13), chronic obstructive pulmonary disease (COPD) (14), and cardiovascular disease (15). Of course, one potentially important difference is that humans have significantly shorter telomeres than mice.

Telomerase Gene Therapy in Adult and Old Mice Delays Aging and Increases Longevity Without Increasing Cancer

In a recent paper Bernardes de Jesus and colleagues suggest that it may be possible to safely express telomerase in adult mice to increase health span and life span (16). They used a non-integrating adeno-associated virus serotype 9 (AAV9) to constitutively express mouse TERT catalytic subunit (mTERT) in a large variety of tissues, including brain and postmitotic cells. Cohorts of 1- and 2-year-old mice were compared. They achieved a sustained (up to 8 months) 5- to 15-fold increase in detectable mTERT expression and at least a 5-fold increase in telomerase activity. AAV9 expressing green fluorescent protein (GFP) was used as a negative control. Bernardes de Jesus observed that, as expected, average telomere length as measured by quantitative telomere FISH increased in treated animals. However, the effect was more modest than might be expected from the strength of the telomerase activity data. In fact, in 1-year old animals, average telomere length only increased between 5% and 10% in most tissues, except in the brain, where no increase was seen, and in the lungs, which saw a 30% increase. In the 2-year-old animals, increases between 10% and 30% were seen, except in muscle, where telomere length was unchanged. On the surface, it appears quite surprising that these modest changes in telomere length would have any phenotypic effects. Therefore, Bernardes de Jesus *et al.* specifically looked at the subset of short telomeres, defined as those that are 15 kb or less, and reported a significant decrease in their number, typically two- to three-fold less for both 1- and 2-year-old mice. One explanation for the differences between the mean and short telomere data might be that telomerase selectively elongates shorter telomeres. Interestingly, "short" 15-kb mouse telomeres are longer than the 10-kb telomeres found in youthful human cells. Thus, the obvious question arises: Why would 15-kb telomeres, which should be quite functional, lead to aging-associated phenotypes?

Nevertheless, animals receiving mTERT showed a significant improvement in several aging-related parameters. For example, age-associated bone loss, which results from osteoblast insufficiency causing bone resorption, as measured by bone mineral density, was completely suppressed by treatment with mTERT in both 1- and 2-year-old mice after 3 months. Loss of the subcutaneous adipose skin layer, which makes the mice more prone to infections, was partially restored in the 2-year-old mice treated with mTERT. Insulin sensitivity generally decreases with age. Treatment of 2-year-old mice with mTERT returned fasting insulin levels to those seen in untreated 1-year old mice, although glucose uptake experiments were less conclusive. Similarly, insulin-like growth factor-1 (IGF-1) levels, which progressively fall during aging and are associated with decreased muscle mass, were restored to normal 1-year-old levels after 2-year-old mice were treated with mTERT. Although, mTERT treatment did not result in statistically improved memory as determined by the object recognition test, coordination and balance as determined by a RotaRod test of 2-year-old mice was comparable to untreated 1 year olds. Neuromuscular coordination, determined by a tightrope test, was seen at levels similar to 1-year-old mice after 11 months of treatment, instead of a 60% decline seen in the untreated 1-year-old animals. The authors suggest that the improvement in coordination and balance, but not memory, may be due to a tropism of AAV9 for motor neurons. Taken together, these results suggest an improvement in health span after mTERT treatment of adult and old mice (16).

Mean life span for the mTERT-treated mice increased 24% for the 1-year-old mice and 13% for the 2-year-old mice. Interestingly, the maximum life spans of the longest-lived mice were 13% greater for the 1-year-old mice and 20% greater for the 2-year-old mice compared to the respective untreated controls. Of paramount importance is that the spontaneous cancer rates did not increase for the mTERT-treated animals, although the 60% cancer rate seen in this mouse strain is already quite high. This is in contrast to the higher cancer rates seen in experiments in which mTERT is expressed in the germ line of wild-type mice (12). The authors suggest that any rapidly dividing cells would quickly lose their extrachromosomal copies of AAV9-TERT, which may prevent such cells from becoming fully tumorigenic. This result raises the possibility that clinical treatment with drugs that induce mTERT may be safer than generally supposed.

One of the keys to clinical development of telomere/telomerase therapeutics is clear understanding of the mechanism of action. There are many possible ways in which telomerase may promote health span and life span, among these increased telomere length and altered intracellular signaling. Modest changes to telomere length were reported (see above), but perhaps not sufficient to explain the health span effects. It is known that TERT can act as a transcriptional activator of the Wnt/beta-catenin signaling pathway (17). Therefore, it is important to determine if Wnt/beta-catenin signaling plays a role in the TERT-mediated effects. Although Bernardes de Jesus did not attempt to block the Wnt/beta-catenin signaling pathway, which could have helped to dissect the mechanism, they investigated expression of downstream Wnt/beta-catenin targets and found that cyclin D1 was stimulated while senescence-associated cyclin-dependent kinase inhibitor p16 was inhibited, as would be expected for ectopic TERT expression. Unfortunately, expression of other TERT/Wnt targets, such as CD44 and axin2, were not consistent with TERT expression and varied by tissue type, perhaps due to tissue-specific WNT/beta-catenin modulators. To determine if stimulation of WNT/beta-catenin alone could result in the health span and life span changes observed, mice were treated with a dominant negative inactive mTERT (dn-mTERT). None of the reported health span or life span changes that were observed with catalytically active mTERT were observed with dn-mTERT.

The authors suggest that this indicates that increased telomere length is the primary effect, but they leave open the possibility that WNT/beta-catenin may play a synergistic role. However, given that transient treatment with the astragalus-derived telomerase inducer TA-65 results in increased telomere length, but not extension of life span (18), one very attractive hypothesis is that both increased telomere length and continuous WNT signaling resulting from the presence of TERT may be required for life span increase. This explanation would go a long way to addressing questions arising from observations that mice have relatively long telomeres, and the authors only observed modest length increases secondary to ectopic telomerase expression. It might also explain any possible effect in post-mitotic cells in which telomere length is expected to be maintained over the life span of the organism. Interestingly, treatment with TA-65 or AAV9-TERT does result in certain improved health span parameters, including glucose tolerance, osteoporosis, and skin fitness, suggesting that increased telomere length alone may be responsible for these effects in both sets of animals.

Medical Implications

Given the current pharmaceutical development environment and safety concerns, it is highly unlikely that systemic AAV9 or similar gene therapy approaches for aging will be developed and approved by the U.S. Food and Drug Administration (FDA) anytime soon. Whether specific organs might be rejuvenated *ex vivo* should be investigated. In the meantime, small-molecule therapeutics that induce human TERT (hTERT) are the best clinical development candidates.

Two small molecules derived from *Astragalus membranaceus*, TA-6519 and TAT2/cycloastragenol (20) have been shown to induce mTERT or hTERT at nanomolar to micromolar doses. TAT2 was shown to retard telomere shortening in human CD8(+) lymphocytes from human immunodeficiency virus (HIV)-infected patients and enhance cytokine/chemo- kine production and antiviral activity (20). In addition to increasing telomere length in mice and increasing health span (18), TA-65 has been shown to moderately activate telomerase in keratinocytes, fibroblasts, and immune cells in preliminary studies in humans (19). Cytomegalovirus (CMV)-seropositive subjects frequently exhibit premature aging of the immune system and present with increased numbers of virus-specific CD8(+)/CD28(-) cytotoxic T and natural killer (NK) cells. Treatment with TA-65 resulted in declines in cytotoxic T and NK cells at 6 and 12 months, which correlates with improved immune function (19).

These promising therapeutics raise multiple questions. First among them is the need for independent replication and confirmation of the findings. Next is a demonstration of safety. On the one hand, two mouse studies, one with TA-65 and one with AAV9-mTERT, showed no increase in numbers of tumors. These studies used either pure C57BL6 mice or mice with a C57BL/6JOlaHsd background. These results must be extended to diverse mouse strains with different spectrums of spontaneous cancers. There may be risks in having very long telomeres, which may afford pre-malignant cells more time to diversify before proliferative crisis or having very short dysfunctional telomeres, which may cause genomic stability, repress immunosurveillance, and promote tumorigenesis (21). Furthermore, does the longer life span of humans present greater danger for oncogenesis than shorter-lived mice?

Another major question relates the mechanism of action. Little is known about mechanism, although inhibitor studies recently implicated activation of c-Src, MEK (ERK kinase), and epidermal growth factor receptor for TAT2/cycloastragenol (22). Given what we now know about telomerase activation and TERT expression, it would be very tempting to postulate that these drugs either directly or indirectly activate the WNT/beta-catenin pathway. That is because not only does TERT stimulate WNT/beta-catenin, but WNT/beta-catenin also activates TERT (23). The reciprocal nature of activation of TERT and WNT/beta-catenin suggests that a potential positive feedback loop may exist for their mutual expression. If so, such a feedback loop might play a significant role in tumorigenesis as well as increase the potential risk of drugs that induce TERT if sustained telomerase expression is not as benign as the recent report from Bernardes de Jesus suggests. Furthermore, the possibility of residual telomerase expression after TA-65 treatment should be investigated.

Conclusions

The possibility of ameliorating a dysfunctional epigenetic change such as telomere shortening has been raised by several studies that show increased health span and life span by expressing TERT in somatic cells in mice. Careful elucidation of the mechanism of action of astragalus-derived drugs may lead to the development of safe therapeutics to delay aging and treat diseases in which telomere shortening plays a critical role. The oncogenic potential of such compounds will need to be thoroughly explored.

References

1. Munoz-Najar U, Sedivy JM. Epigenetic control of aging. Antioxid Redox Signal 2011;14:241–259.
2. Tan TCJ, Rahman R, Jaber-Hijazi F, Felix DA, Chen C, Louis EJ, Aboobaker A. Telomere maintenance and telomerase activity are differentially regulated in asexual and sexual worms. Proc Natl Acad Sci USA 2012;109:4209–4214.
3. Tomas-Loba A, Flores I, Fernandez-Marcos PJ, Cayuela ML, Maraver A, Tejera A, Borras C, Matheu A, Klatt P, Flores JM, Vina J, Serrano M, Blasco MA. Telomerase reverse transcriptase delays aging in cancer-resistant mice. Cell 2008;135:609–622.
4. Artandi SE, Alson S, Tietze MK, Sharpless NE, Ye S, Greenberg RA, Castrillon DH, Horner JW, Weiler SR, Carrasco RD, DePinho RA. Constitutive telomerase expression promotes mammary carcinomas in aging mice. Proc Natl Acad Sci USA 2002;99:8191–8196.

5. Canela A, Martın-Caballero J, Flores JM, Blasco MA. Constitutive expression of TERT in thymocytes leads to increased incidence and dissemination of T-cell lymphoma in LCK-TERT mice. Mol Cell Biol 2004;24:4275–4293.

6. Gonzalez-Suarez E, Samper E, Ramirez A, Flores JM, Martn-Caballero J,Jorcano JL, Blasco MA. Increased epidermal tumors and increased skin wound healing in transgenic mice overexpressing the catalytic subunit of telomerase, mTERT, in basal keratinocytes. EMBO J 2001;20:2619–2630.

7. McKay JD, Hung RJ, Gaborieau V, Boffetta P, Chabrier A, Byrnes G, Zaridze D, Mukeria A, Szeszenia-Dabrowska N, Lissowska J, Rudnai P, Fabianova E, Mates D, Bencko V, Foretova L, Janout V, McLaughlin J, Shepherd F, Montpetit A, Narod S, Krokan HE, Skorpen F, Elvestad MB, Vatten L, Njølstad I, Axelsson T, Chen C, Goodman G, Barnett M, Loomis MM, Lubinski J, Matyjasik J, Lener M, Oszutowska D, Field J, Liloglou T, Xinarianos G, Cassidy A, Zelenika D, Boland A, Delepine M, Foglio M, Lechner D, Matsuda F, Blanche H, Gut I, Heath S, Lathrop M, Brennan P, Vineis P, Clavel-Chapelon F, Palli D, Tumino R, Krogh V, Panico S, Gonzalez CA, Quiros JR, Martınez C, Navarro C, Ardanaz E, Larranaga N, Kham KT, Key T, Bueno-de-Mesquita HB, Peeters PHM, Trichopoulou A, Linseisen J, Boeing H, Hallmans G, Overvad K, Tjønneland A, Kumle M, Riboli E. Lung cancer susceptibility locus at 5p15(3)3. Nat Genet 2008;40:1404–1406.

8. Rafnar T, Sulem P, Stacey SN, Geller F, Gudmundsson J, Sigurdsson A, Jakobsdottir M, Helgadottir H, Thorlacius S, Aben KKH, Blondel T, Thorgeirsson TE, Thorleifsson G, Kristjansson K, Thorisdottir K, Ragnarsson R, Sigurgeirsson B, Skuladottir H, Gudbjartsson T, Isaksson HJ, Einarsson GV, Benediktsdottir KR, Agnarsson BA, Olafsson K, Salvarsdottir A, Bjarnason H, Asgeirsdottir M, Kristinsson KT, Matthiasdottir S, Sveinsdottir SG, Polidoro S, Hoiom V, Botella-Estrada R, Hemminki K, Rudnai P, Bishop DT, Campagna M, Kellen E, Zeegers MP, de Verdier P, Ferrer A, Isla D, Vidal MJ, Andres R, Saez B, Juberias P, Banzo J, Navarrete S, Tres A, Kan D, Lindblom A, Gurzau E, Koppova K, de Vegt F, Schalken JA, van der Heijden HFM, Smit HJ, Termeer RA, Oosterwijk E, van Hooij O, Nagore E, Porru S, Steineck G, Hansson J, Buntinx F, Catalona WJ, Matullo G, Vineis P, Kiltie AE, Mayordomo JI, Kumar R, Kiemeney LA, Frigge ML, Jonsson T, Saemundsson H, Barkardottir RB, Jonsson E, Jonsson S, Olafsson JH, Gulcher JR, Masson G, Gudbjartsson DF, Kong A, Thorsteinsdottir U, Stefansson K. Sequence variants at the tert-clptm1l locus associate with many cancer types. Nat Genet 2009;41:221–227.

9. Sahin E, Colla S, Liesa M, Moslehi J, Mu ̈ller FL, Guo M, Cooper M, Kotton D, Fabian AJ, Walkey C, Maser RS, Tonon G, Foerster F, Xiong R, Wang YA, Shukla SA, Jaskelioff M, Martin ES, Heffernan TP, Protopopov A, Ivanova E, Mahoney JE, Kost-Alimova M, Perry SR, Bronson R, Liao R, Mulligan R, Shirihai OS, Chin L, DePinho RA. Telomere dysfunction induces metabolic and mitochondrial compromise. Nature 2011;470:359–365.

10. Rudolph KL, Chang S, Lee HW, Blasco M, Gottlieb GJ, Greider C, DePinho

RA. Longevity, stress response, and cancer in aging telomerase-deficient mice. Cell 1999;96:701–712.

11. Begus-Nahrmann Y, Hartmann D, Kraus J, Eshraghi P, Scheffold A, Grieb M, Rasche V, Schirmacher P, Lee H-W, Kestler HA, Lechel A, Rudolph KL. Transient telomere dysfunction induces chromosomal instability and promotes carcinogenesis. J Clin Invest 2012;122:2283–2288.

12. Jaskelioff M, Muller FL, Paik J-H, Thomas E, Jiang S, Adams A, Sahin E, Kost-Alimova M, Protopopov A, Cadinanos J, Horner JW, Maratos-Flier E, DePinho RA. Telomerase reactivation reverses tissue degeneration in aged telomerase deficient mice. Nature 2011;469:102–106.

13. Georgin-Lavialle S, Aouba A, Mouthon L, Londono-Vallejo JA, Lepelletier Y, Gabet A-S, Hermine O. The telomere/ telomerase system in autoimmune and systemic immune-mediated diseases. Autoimmun Rev 2010;9:646–651.

14. Tsuji T, Aoshiba K, Nagai A. Alveolar cell senescence in patients with pulmonary emphysema. Am J Respir Crit Care Med 2006;174:886–893.

15. Harst P van der, Veldhuisen DJ van, Samani NJ. Expanding the concept of telomere dysfunction in cardiovascular disease. Arterioscler Thromb Vasc Biol 2008;28:807–808.

16. Bernardes de Jesus B, Vera E, Schneeberger K, Tejera AM, Ayuso E, Bosch F, Blasco MA. Telomerase gene therapy in adult and old mice delays aging and increases longevity without increasing cancer. EMBO Mol Med 2012; first published online May 15, 2012, DOI: 10(1)002/emmm.201200245

17. Park J-I, Venteicher AS, Hong JY, Choi J, Jun S, Shkreli M, Chang W, Meng Z, Cheung P, Ji H, McLaughlin M, Veenstra TD, Nusse R, McCrea PD, Artandi SE. Telomerase modulates wnt signalling by association with target gene chromatin. Nature 2009;460:66–72.

18. de Jesus BB, Schneeberger K, Vera E, Tejera A, Harley CB, Blasco MA. The telomerase activator TA-65 elongates short telomeres and increases health span of adult/old mice without increasing cancer incidence. Aging Cell 2011;10:604–621.

19. Harley CB, Liu W, Blasco M, Vera E, Andrews WH, Briggs LA, Raffaele JM. A natural product telomerase activator as part of a health maintenance program. Rejuvenation Res 2011;14:45–56.

20. Fauce SR, Jamieson BD, Chin AC, Mitsuyasu RT, Parish ST, Ng HL, Kitchen CMR, Yang OO, Harley CB, Effros RB. Telomerase-based pharmacologic enhancement of antiviral function of human CD8+ T lymphocytes. J Immunol 2008;181:7400–7406.

21. Baird DM. Variation at the TRET locus and predisposition for cancer. Expert Rev Mol Med 2010;18:e16.

22. Yung LY, Lam WS, Ho MKC, Hu Y, Ip FCF, Pang H, Chin AC, Harley CB, Ip NY, Wong YH. Astragaloside iv and cycloastragenol stimulate the phosphorylation of extracellular signal-regulated protein kinase in multiple cell types. Planta Med

2012;78:115–121.
23. Hoffmeyer K, Raggioli A, Rudloff S, Anton R, Hierholzer A, Del Valle I, Hein K, Vogt R, Kemler R. Wnt/beta-catenin signaling regulates telomerase in stem cells and cancer cells. Science 2012;336:1549–1554.

Section III

Modulating metabolism and growth regulation

Chapter 4

Rapamycin as an anti-aging therapeutic?: targeting mammalian target of rapamycin to treat Hutchinson-Gilford progeria and neurodegenerative diseases

Mammalian target of rapamycin (mTOR), a serine/threonine kinase and component of the mTORC1 signaling complex, acts as an energy, nutrient, growth factor, stress, and redox sensor to increase protein synthesis and decrease macroautophagy. mTORC1 plays a central role in the maintenance of homeostasis and its deterioration, seen in aging. The Food and Drug Administration (FDA)-approved immunosuppressive macrolide rapamycin binds immunophilin FKBP12 (FK506-binding protein) to inhibit mTORC1. Unlike most other interventions tested to date, inhibition of mTORC1 by rapamycin extends life span in old mice, likely by a combination of increased autophagy and decreased mRNA translation. Hutchinson–Gilford progeria syndrome (HGPS) is a lethal genetic disorder affecting children that is characterized by symptoms of premature aging, such as atherosclerosis. Increased autophagy induced by rapamycin reduces accumulation of progerin, an alternate spliced form of lamin A/C, that forms insoluble toxic aggregates, resulting in reduced HGPS-associated nuclear blebbing, growth inhibition, epigenetic dysregulation, and genomic instability. Rapamycin-induced autophagy also suppresses symptoms in mouse models of Alzheimer's, Parkinson's, and Huntington's diseases, where toxic insoluble protein aggregates accumulate. On the basis of these results, modulation of mTORC1 function is a promising target for the development of therapeutics for neurodegenerative diseases and HGPS. Rapamycin is the obvious candidate for near-term evaluation in the treatment of these diseases. However, the substantial set of rapamycin-associated adverse effects, as well as the lack of aging-specific human data, should caution the routine use of rapamycin as an anti-aging agent. The use of safer, but perhaps weaker, indirect mTORC1 inhibitors, such as metformin and resveratrol, may prove useful. Further study will ascertain whether such compounds extend human health or life span.

Inhibition of Mammalian Target of Rapamycin Extends Life and Health Span

The search for interventions to extend health and life span has identified the evolutionarily conserved, metazoan target of rapamycin (TOR) pathway as a possible target. Inhibition of the TOR pathway extends health and life span in a variety of model organisms, including *Saccharomyces cerevisiae*, *Caenorhabditis elegans*, *Drosophila melanogaster*, and *Mus musculus* (1,2). The TOR pathway plays a major role in maintaining homeostasis by integrating nutrient and growth factor status with cellular metabolism. Because aging involves dysregulation of homeostasis, it is reasonable to expect that alteration of a major hub of growth signaling might influence the rate of aging. However, TOR signaling has been incorporated into diverse signaling pathways beyond simple growth control and autophagy (3) which in mammals include control of the adaptive immune response (3,4) and neurological (5) memory formation-consolidation. This complicates the potential use of TOR inhibitors as modulators of aging as well as efforts to decipher the molecular mechanisms by which TOR inhibition extends life span.

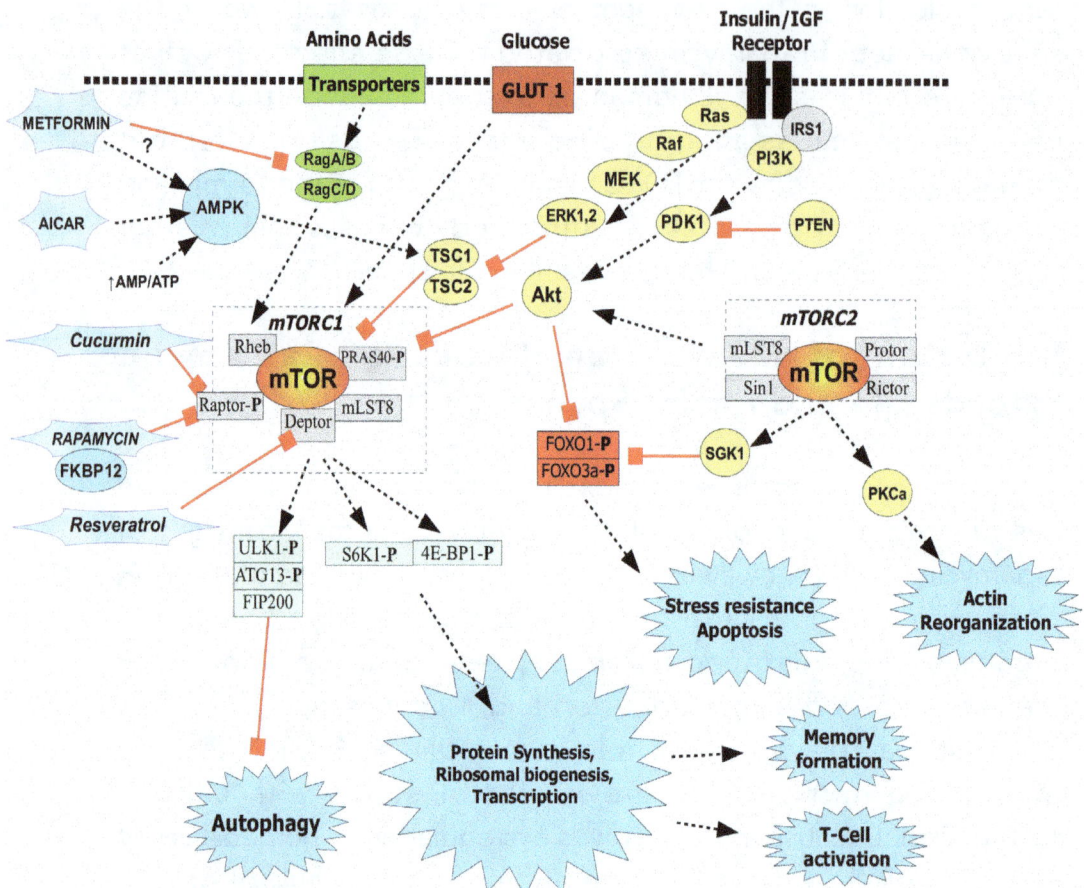

Figure 4.1 Simplified schematic of mammalian target of rapamycin (mTOR) signaling pathways. mTORC1 promotes messenger RNA (mRNA) translation and inhibits autophagy by integrating growth factor signals like insulin/inslin-like growth factor (IGF), amino acid nutrient signals, and adenosine triphosphate (ATP) levels via 5′-adenosine monophosphate-activated protein kinase (AMPK). mTORC1 also senses DNA damage and hypoxia (not shown). mTORC1 affects many homeostatic processes beyond cell growth and autophagy, including neural memory formation/consolidation and immunity. Growth factors activate mTORC2 to activate Akt and Sgk1, which promote survival and inhibit forkhead transcription factors FOXO1 and FOXO3. The forkhead transcription factors promote antioxidant enzyme expression and cell death. The molecular targets of mTORC1 inhibitors, rapamycin, curcumin, resveratrol, metformin, and 5-aminoimidazole-4-carboxamide-1-β-d-ribofuranoside (AICAR), are identified.

The mammalian TOR (mTOR), a serine/threonine kinase, is found in two signaling complexes, mTORC1 and mTORC2 (see **Figure 4.1**). mTORC1 acts as an energy, nutrient, growth factor, mitogen, stress, and redox sensor and controls protein synthesis and autophagy in a reciprocal fashion (3). For example, mTORC1 is activated by increased energy and nutrients to positively regulate S6 kinase (S6K) and eIF4e-BP(4E-BP1) to initiate and augment translation. Autophagasomal regulators Unc51-like kinase 1 (ULK1) and ATG13 are inhibited by mTORC1 phosphorylation. Although less well characterized, mTORC2 responds to growth factors to regulate cell polarization via the actin cytoskeleton and cell survival and metabolism via phosphorylation of pro-survival Akt, protein kinase C, and glucocorticoid-regulated kinase (SGK) (4). Akt promotes survival by phosphorylating proapoptotic transcription factors FOXO1 and FOXO3 to inhibit their translocation to the nucleus. Activation of mTORC2 can also contribute to increased activity of mTORC1 via Akt activation of TSC2 and Rheb4 (see **Figure 4.1**). The immunosuppressive macrolide rapamycin, which binds the immunophilin FK506- binding protein (FKBP12), specifically inhibits mTORC1, but not mTORC2 by interfering with the interaction of mTOR with the mTORC1-specific Raptor protein(3,4), Unfortunately, this picture is muddied in that chronic high doses of rapamycin can also inhibit mTORC2 in some cell types (6).

Unlike most other interventions, inhibition of mTORC1 by rapamycin extends life span in mice, even when given to old individuals (600 days old) (7). The mechanisms that underlie this intriguing result are under investigation. However, there are two non–mutually exclusive mechanisms that are thought to play an important role, increased autophagy8 and decreased mRNA translation. Overexpression of Atg8 is known to increase autophagy and extend life span in fruit flies (2). However, in other model organisms (worms and mice), increased autophagy does not extend life span but appears to be necessary for life extension by other means, such as calorie restriction (CR) and rapamycin. Optimal levels of autophagy are undoubtedly important because high levels can contribute to cell death. In contrast to the autophagy results, reduced translation via knockout of S6 protein kinase 1 (S6PK1) in mice contributes to increased life span and increased resistance to age-related pathologies, such as loss of insulin sensitivity and motor, bone, and immune dysfunction.9 It is not unreasonable to hypothesize that both mechanisms play an important role in extension of longevity in mammals (2).

Rapamycin Suppresses Hutchinson–Gilford Progeria Syndrome in Tissue Culture

In a fascinating study, Cao *et al* (10) show that rapamycin can suppress the phenotype of Hutchinson–Gilford progeria syndrome (HGPS) cells in culture. HGPS is a lethal genetic disorder affecting children who exhibit many characteristics of premature aging, including alopecia, osteoporosis, sclerodermatous skin, and atherosclerosis. Patients with HGPS have an average lifespan of 12 years and typically die from myocardial infarction or stroke. HGPS is caused by a genetic mutation that partially activates a cryptic splice site in exon 11 of the lamin A/C gene, which encodes a nuclear envelope protein. The resulting abnormal protein, progerin, is thought to cause HGPS (11). Interestingly, progerin is expressed to a small extent in wild-type humans and slowly accumulates with age. Primary fibroblasts from HGPS patients exhibit a variety of abnormal phenotypes, including nuclear blebbing, punctate cytoplasmic accumulation of progerin (12), reduced growth rate, and decreased Hayflick limit. As might be expected from a protein intimately associated with nuclear function, significant epigenetic changes are associated with HGPS, particularly alterations in histone methylation. Aberrant epigenetic regulation (developmental drift) (13) and accumulation of aggregated proteins (14) are under active investigation as key processes by which cells age. Such work reinforces the relevance of HGPS as a model of premature aging as well as an important syndrome in need of new therapies (15).

Treatment of HGPS cells for 10 days with rapamycin reduced nuclear blebbing, as determined morphologically by anti-tubulin staining. In addition, the reduced cellular growth rate associated with HGPS was reversed by rapamycin. This was somewhat surprising to the authors because rapamycin itself is known to inhibit cell proliferation. However, the ability of rapamycin to inhibit the rate of cellular growth varies as some cells respond differentially to lower S6K and eIF4e-BP activity (16). Rapamcyin also delayed replicative senescence by 30 days (& 10 passages) in both normal and HGPS cells, as assessed by senescence-associated beta-galactosidase expression and cell number (10).

Accumulating evidence suggests that epigenetic dysregulation plays a major role in aging (7). Progerin has been shown to progressively alter trimethylation of histone H3 Lys27 (H3K27me3), which marks tightly packed, transcriptionally inactive, facultative heterochromatin (17). In normal female cells, H3K27me3 coats the inactivated X chromosome (Xi). However, in female HGPS cells, coating of Xi by H3K27me3 is partially lost and is restored by exposure to rapamycin (10).

Genomic instability has also been hypothesized to play a role in aging. HGPS cells exhibit genomic instability, including increased sensitivity to DNA cross-linker mitomycin C and increased expression of DNA-damage associated p53-binding protein 1 (TP53BP1)(1)(8) Rapamycin treatment of HGPS cells significantly reduced anti-TP53BP1 nuclear staining as well as cellular sensitivity to mitomycin C (10). Because HGPS results from accumulation of progerin, which appears to correlate with nuclear blebbing (12), the authors hypothesized that rapamycin would lower levels of progerin by increasing its degradation via autophagy. Indeed, rapamycin was seen to reduce nuclear progerin in HGPS nuclei, as well as the half-life of progerin measured by a pulse-chase experiment. Two experiments supported the beneficial role of rapamycin to reduce progerin levels via increased autophagy: (1) Bafilomycin A, a lysosomal pump inhibitor reversed the effect in HPGS cells; and (2) small interfering RNA (siRNA) to ATG7 (known to be essential for autophagy) reversed the effect in a HeLa cell-based progerin expression model. Progerin was found to interact with autophagy adapter protein p62 through K63 ubiquitination and autophagy-linked zinc finger protein FYVE, which functionally links progerin with autophagasomes (10).

Age-associated increases in protein aggregation is another mechanism that may play a major role in aging (19). Progerin has been shown to aggregate with lamin A to form insoluble complexes that may promote cellular dysfunction. Rapamycin treatment increased the solubility of progerin, by promoting formation of a soluble lamin A progerin A dimer (10) This may result from the reduced progerin levels found in rapamycin-treated cells. In summary, Cao *et al.* suggest a central role of augmented autophagy in the beneficial effects of rapamycin, although the potential role of decreased initiation and elongation of translation was not well explored. Cao *et al.* use these basic science findings to make a strong case in favor of the initiation of clinical trials for treatment of inevitably fatal HGPS.

Rapamycin Suppresses Neurodegenerative Diseases in Mouse Models

Given the beneficial effects of rapamycin on HGPS cells, it would be expected that rapamycin might also benefit other diseases in which clearance of toxic protein aggregates is helpful. Indeed, rapamycin has been shown to inhibit pathology in mouse models of neurodegenerative diseases in which protein aggregation plays a pathological role: Alzheimer's, Parkinson's, and Huntington's diseases (20–22). The two mouse models of Alzheimer's disease are triple transgenic mice with mutations in PS1, amyloid precursor protein (APP), and tau, and the PDAPP mouse, which carries a mutant APP, rapamycin-extended life span, and reduced beta-amyloid levels, plaques, and neurofibrillary tangles. Importantly, the levels of rapamycin used did not impair memory (20) (see below).

In a 1-methyl-4-phenyl-1,2,3,6-tetrahydro-pyridine (MPTP)-induced Parkinson's disease mouse model, rapamycin helped maintain dopaminergic neuron cell numbers and reduced loss of 3,4-dihydroxyphenylacetic acid (DOPAC), a metabolite of dopamine. The authors hypothesize that these effects are mediated by activation of autophagic/lysosomal pathways (21). In a mouse model of Huntington's disease, rapamycin reduces symptoms of polyglutamine neurotoxicity. Ross/Borchelt mice expressing huntingtin with 82 glutamine repeats treated with rapamycin before the onset of symptoms exhibited reduced numbers of inclusions and improved motor function (22).

Rapamycin may not be altogether beneficial to brain function. mTORC1 plays a key role in synaptic plasticity by modulating protein synthesis. Late-phase synaptic plasticity is known to be attenuated by rapamycin. Rapamycin has also been shown to disrupt the transition of short-term to long-term memory and interregional memory consolidation in mouse models (20). However, subtle variation in mTOR activity is likely important, because increased mTORC1 activity also can disrupt memory processing. In fact, in a mouse model of tuberous sclerosis, a disease affecting multiple tissues that causes tuber- or root-shaped growths in the brain, rapamycin treatment rescued memory deficits. Because inhibition of mTOR has been seen to be a possible dietary restriction (DR) mimetic and because DR increases neurogenesis and reduces the rate of aging-dependent brain deterioration and memory loss in rodents, it might be predicted that rapamycin would have similar benefit in humans. Retrospective memory studies on patients treated with rapamycin for other indications, such as suppression of organ transplantation over long periods of time, might shed light on the usefulness of rapamycin in neurodegeneration, as well as any memory-related side-effects.

Rapamycin and Other Compounds That Target mTORC1 As Antiaging Therapeutics

Inhibition of mTORC1 by rapamycin has substantial beneficial effects in mouse models, including increased longevity and inhibition of engineered neurodegenerative diseases. Intriguing data suggest that rapamycin is also able to suppress HGPS in cultured human cells, making it a potential therapeutic for HGPS as well as neurodegenerative diseases that involve accumulation of toxic protein aggregates. Is there an additional role for mTORC1 inhibitors such as rapamycin to retard aging?

Rapamycin and related compounds carry substantial risk of adverse side effects, which include hypertriglyceridemia (57%), hypercholesterolemia (46%), arthralgia (31%), peripheral edema (58%), fever (34%), pain (33%), anemia (33%), thrombocytopenia (30%), constipation (38%), abdominal pain (36%), diarrhea (35%), nausea (31%), hypertension (49%), and decreased wound healing (www.drugs.com/ ppa/sirolimus.html). Some of these risks may be addressed by combination with other agents. For example, hyper-triglyceridemia can be relieved by co-administration of nutritional supplement omega-3 fatty acid (23). Perhaps the most serious common side effects are mucositis and pneumonitis, which can be medically managed (24).

Additional serious potential side effects of rapamycin due to its ability to induce immunosuppression, such as increased rate of infections and cancer, are understudied and probably exaggerated. In fact, there are preliminary clinical data indicating that mTORC1 inhibitors such as rapamycin or everolimus may actually be useful in treating many types of cancers, including relapsed aggressive lymphoma (25), mantle-cell lymphoma, tumors associated with treatment with cyclosporine, an immunosupressor (26), neuroendocrine tumors, and metastatic renal carcinoma (27). Paradoxically, rapamycin actually enhances resistance to some viral infections, such as transplantation-associated late-stage cytomegalovirus (28). It is possible that the general antiproliferative effects of mTORC1 inhibition on cancer cells outweigh possible reduction in immune surveillance due to reduced Th1 (cell-mediated immunity) and Th17 (tissue lining protection) T cell differentiation and increased maintenance of tolergenic dendritic cells (29). For infectious diseases, reduced numbers of Th1 and Th17 cells may be counterbalanced by increased antigen presentation due to increased autophagy(30). It is also entirely possible that rapamycin does not cause more serious side effects because rapamycin-mediated inhibition of mTORC1 is incomplete (31).

The safety profile of rapamycin and related compounds should give pause to its consideration as as a potential anti-aging retardant in the absence of potentially life-threatening pathologies. However, there are other compounds that indirectly target mTORC1 with better safety profiles. For example, the antidiabetic drug metformin inhibits mTORC1 by inhibiting RAG guanosine triphosphatases (GTPases) that would otherwise stimulate mTORC1(32). Metformin and 5-aminoimidazole-4-carboxamide-1-b-d-ribofuranoside (AICAR) activate 5'-adenosine monophosphate (AMP) kinase, which has been reported to inhibit mTORC1 indirectly and perhaps only weakly (32). Curcumin may disrupt the interaction of Raptor with mTORC1(33). Resveratrol can inhibit mTORC1 by blocking the interaction between Deptor and mTOR (34). Ultimately, retardation of aging via inhibition of mTORC1 may depend on the mechanism of inhibition and possible interactions with other targets that attenuate potential anti-aging effects. For example, rapamycin extends murine life span, whereas to date resveratrol has not extended life span in wild-type mice fed a normal diet.

Conclusions

mTORC1 is a promising target for the development of therapeutics for neurodegenerative diseases and HGPS. Rapamycin, an Food and Drug Administration (FDA)-approved drug used to induce immunological tolerance, should be tested for clinical relevance in the treatment of these diseases. However, the substantial set of rapamycin-associated side effects, as well as the lack of aging-specific human data, should caution those who wish to use rapamycin or closely related compounds to slow aging. The use of safer, but perhaps weaker, indirect mTORC1 inhibitors, such as metformin and resveratrol, may prove advantageous. Further study will be useful in ascertaining whether such compounds extend human health or life span.

References

1. Bjedov I, Partridge L. A longer and healthier life with tor down-regulation: Genetics and drugs. Biochem Soc Trans 2011;39:460–465.
2. Hands SL, Proud CG, Wyttenbach A. mTOR's role in ageing: Protein synthesis or autophagy? Aging (Albany NY) n.d(1):586–597.
3. Laplante M, Sabatini DM. mTOR signaling at a glance. J Cell Sci 2009;122:3589–3594.
4. Zoncu R, Efeyan A, Sabatini DM. mTOR: From growth signal integration to cancer, diabetes and ageing. Nat Rev Mol Cell Biol 2011;12:21–35.
5. Bekinschtein P, Katche C, Slipczuk LN, Igaz LM, Cammarota M, Izquierdo I, Medina JH. mTOR signaling in the hippocampus is necessary for memory formation. Neurobiol Learn Mem 2007;87:303–307.
6. Foster DA, Toschi A. Targeting mTOR with rapamycin: One dose does not fit all. Cell Cycle 2009;8:1026–1029.
7. Harrison DE, Strong R, Sharp ZD, Nelson JF, Astle CM, Flurkey K, Nadon NL, Wilkinson JE, Frenkel K, Carter CS, Pahor M, Javors MA, Fernandez E, Miller RA. Rapamycin fed late in life extends lifespan in genetically heterogeneous mice. Nature 2009;460:392–395.
8. Madeo F, Tavernarakis N, Kroemer G. Can autophagy promote longevity? Nat Cell Biol 2010;12:842–846.
9. Selman C, Tullet JMA, Wieser D, Irvine E, Lingard SJ, Choudhury AI, Claret M, Al-Qassab H, Carmignac D, Ramadani F, Woods A, Robinson ICA, Schuster E,

Batterham RL, Kozma SC, Thomas G, Carling D, Okkenhaug K, Thornton JM, Partridge L, Gems D, Withers DJ. Ribosomal protein s6 kinase 1 signaling regulates mammalian life span. Science 2009;326:140–144.

10. Cao K, Graziotto JJ, Blair CD, Mazzulli JR, Erdos MR, Krainc D, Collins FS. Rapamycin reverses cellular phenotypes and enhances mutant protein clearance in Hutchinson-Gilford progeria syndrome cells. Sci Transl Med 2011;3:89ra58.

11. Eriksson M, Brown WT, Gordon LB, Glynn MW, Singer J, Scott L, Erdos MR, Robbins CM, Moses TY, Berglund P, Dutra A, Pak E, Durkin S, Csoka AB, Boehnke M, Glover TW, Collins FS. Recurrent de novo point mutations in lamin A cause Hutchinson-Gilford progeria syndrome. Nature 2003;423:293–298.

12. Goldman RD, Shumaker DK, Erdos MR, Eriksson M, Goldman AE, Gordon LB, Gruenbaum Y, Khuon S, Mendez M, Varga R, Collins FS. Accumulation of mutant lamin a causes progressive changes in nuclear architecture in Hutchinson-Gilford progeria syndrome. Proc Natl Acad Sci USA 2004;101:8963–8968.

13. Budovskaya YV, Wu K, Southworth LK, Jiang M, Tedesco P, Johnson TE, Kim SK. An elt3/elt-5/elt-6 gata transcription circuit guides aging in *C. elegans*. Cell 2008;134:291–303.

14. Mendelsohn A, Larrick JW. Medical implications of basic science: Protein homeostasis as a clinical target for increased longevity? Rejuvenation Res 2011;14:335–339.

15. Burtner CR, Kennedy BK. Progeria syndromes and ageing: What is the connection? Nat Rev Mol Cell Biol 2010;11: 567–578.

16. Gruppuso PA, Boylan JM, Sanders JA. The physiology and pathophysiology of rapamycin resistance: Implications for cancer. Cell Cycle 2011;10:1050–1058.

17. Shumaker DK, Dechat T, Kohlmaier A, Adam SA, Bozovsky MR, Erdos MR, Eriksson M, Goldman AE, Khuon S, Collins FS, Jenuwein T, Goldman RD. Mutant nuclear lamin A leads to progressive alterations of epigenetic control in premature aging. Proc Natl Acad Sci USA 2006;103:8703–8708.

18. Liu B, Wang J, Chan KM, Tjia WM, Deng W, Guan X, Huang J-dong, Li KM, Chau PY, Chen DJ, Pei D, Pendas AM, Cadiûanos J, Lù pez-OtÆn C, Tse HF, Hutchison C, Chen J, Cao Y, Cheah KSE, Tryggvason K, Zhou Z. Genomic instability in laminopathy-based premature aging. Nat Med 2005;11: 780–785.

19. David DC, Ollikainen N, Trinidad JC, Cary MP, Burlingame AL, Kenyon C. Widespread protein aggregation as an inherent part of aging in *C. elegans*. PLoS Biol 2010;8: e1000450.

20. Garelick MG, Kennedy BK. TOR on the brain. Exp Gerontol 2011;46:155163.

21. Liu K,Liu C, Shen L, Shi J, Zhang T, Zhou Y ,Zhou L, Sun X, . Therapeutic effects of rapamycin on MPTP-induced parkinsonism in mice. Neurochem Int 2011;doi:10(1)016/j.neuint .2011.05.011. 22. Ravikumar B, Vacher C, Berger Z, Davies JE, Luo S, Oroz LG, Scaravilli F, Easton DF, Duden R, O'Kane CJ, Rubinsztein DC. Inhibition of mTOR induces autophagy and reduces toxicity of

polyglutamine expansions in fly and mouse models of Huntington's disease. Nat Genet 2004;36: 585–595.
23. Celik S, Doesch A, Erbel C, Blessing E, Ammon K, Koch A, Katus HA, Dengler TJ. Beneficial effect of omega-3 fatty acids on sirolimus- or everolimus-induced hypertriglyceridemia in heart transplant recipients. Transplantation 2008; 86:245–250.
24. Sankhala K, Mita A, Kelly K, Mahalingam D, Giles F, Mita M. The emerging safety profile of mTOR inhibitors, a novel class of anticancer agents. Target Oncol 2009;4:135–142.
25. Witzig TE, Reeder CB, LaPlant BR, Gupta M, Johnston PB, Micallef IN, Porrata LF, Ansell SM, Colgan JP, Jacobsen ED, Ghobrial IM, Habermann TM. A phase II trial of the oral mTOR inhibitor everolimus in relapsed aggressive lymphoma. Leukemia 2011;25:341–347.
26. Leblanc KG Jr, Hughes MP, Sheehan DJ. The role of sirolimus in the prevention of cutaneous squamous cell carcinoma in organ transplant recipients. Dermatol Surg 2011;37: 744–749.
27. Dancey J. mTOR signaling and drug development in cancer. Nat Rev Clin Oncol 2010;7:209–219.
28. Cervera C, Fernandez-Ruiz M, Valledor A, Linares L, Anton A, Angeles Marcos M, Sanclemente G, Hoyo I, Cofan F, Ricart MJ, Parez-Villa F, Navasa M, Pumarola T, Moreno A. Epidemiology and risk factors for late infection in solid organ transplant recipients. Transpl Infect Dis 2011;DOI:10(1)111/j(1)399-3062.2011.00646.x.
29. Salmond RJ, Zamoyska R. The influence of mTOR onT helper cell differentiation and dendritic cell function. Eur J Immunol 2011;41:2137–2141.
30. Peter C, Waldmann H, Cobbold SP. mTOR signalling and metabolic regulation of T cell differentiation. Curr Opin Immunol 2010;22:655–661.
31. Thoreen CC, Sabatini DM. Rapamycin inhibits mTORC1, but not completely. Autophagy 2009;5:725–726.
32. Kalender A, Selvaraj A, Kim SY, Gulati P, BrulA SB, Viollet B, Kemp B, Bardeesy N, Dennis P, Schlager JJ, Marette A, Kozma SC, Thomas G. Metformin, independent of AMPK, inhibits mTORC1 in a rag GTPase-dependent manner. Cell Metab 2010;11:390–401.
33. Beevers CS, Chen L, Liu L, Luo Y, Webster NJG, Huang S. Curcumin disrupts the mammalian target of rapamycin-raptor complex. Cancer Res 2009;69:1000–1008.
34. Liu M, Wilk SA, Wang A, Zhou L, Wang R-H, Ogawa W, Deng C, Dong LQ, Liu F. Resveratrol inhibits mTOR signaling by promoting the interaction between mTOR and DEPTOR. J Biol Chem 2010;285:36387–36394.

Chapter 5

Reversing Age-Related Decline in Working Memory

Higher cognitive functions, such as working memory and the ability to focus attention, decline as people age. Recently, it has been reported that decline in working memory in aging rhesus monkeys correlates with the loss of activity of a specific set of neurons in the prefrontal cortex during a delay following a learning cue. The activity of these neurons can be rescued by stimulating alpha-2 adrenergic receptors, inhibiting cyclic adenosine monophosphate (cAMP) signaling, or closing potassium channels that are known to inhibit firing and synaptic connectivity. Agents that stimulate neurons expressing a-2 adrenergic receptors may prove useful in treating working memory loss in humans.

Introduction

Higher cognitive functions decline as humans age, even in the absence of neurodegenerative disease. Cognitive flexibility, task set maintenance, divided attention, and working memory are among those processes that are the first to decrease during aging, beginning in middle age (1,2). Working memory changes are exemplified by gradual loss of ability to remember recent events (What was the name of the person who just introduced herself?) or to recall information from long-term memory (Where did I put my keys last night?). Working memory acts as a critical cog in cognition. The prefrontal cortex, especially area 46 of the dorsolateral prefrontal cortex, plays a critical role in mediating these executive functions (3). Continual self-excitation of a network of pyramidal neurons in area 46 maintains working memory after an initial stimulatory event has ended (3). Although, various age-associated changes have been observed in area 46, including reduced volume, synapse density, micro- column strength, and alpha 1- and alpha 2- adrenergic receptor numbers, neuronal loss does not occur (3). The specific mechanisms by which executive function is gradually lost with aging in the prefrontal cortex have not been clearly identified, making successful intervention difficult.

Neuronal Basis of Age-Related Working Memory Decline

In an important paper, Wang and colleagues (4) identify specific physiological changes that occur in neurons that maintain working memory in the prefrontal cortex of rhesus monkeys (*Macaca mulatta*). Just as in humans, working memory in *M. mulatta* depends on area 46 of the prefrontal cortex, and just as in humans, working memory declines in middle age (these monkeys age around three times faster than humans). Importantly, unlike humans, rhesus monkeys do not suffer from neurodegenerative diseases such as Alzheimer's disease, so deficits resulting from primary aging mechanisms can be studied (3).

Three groups of rhesus monkeys (young, middle-aged, and old) were trained to remember a spatial location. They were first shown a cue, then after a brief delay period of 2.5 sec in which no spatial information was present, they were prompted by the disappearance of a fixation spot to move their eyes to the remembered location for a juice reward—the response period. The spatial location changed randomly on each trial. At the same time, neural activity was recorded from individual neurons from area 46. CUE neurons fire only during the cue period; DELAY neurons fire during the critical delay period as well as during the cue and response periods. The DELAY neurons are required for working memory. Many of the DELAY neurons fire only when presented with the memory of one specific spatial location, but not other "anti-preferred" locations, 180 degrees opposite to the preferred direction (4).

With advancing age, there are strong reductions in spontaneous and task-related firing of DELAY neurons. The reduction of activity is particularly notable in the preferred direction and less so in the anti-preferred direction. This causes a decreased difference in the firing of DELAY neurons in the preferred versus anti-preferred directions with advancing age. This difference represents a diminished ability to distinguish the preferred from the non-preferred location during the delay period, in other words, a diminished ability to maintain spatial working memory. Neural circuit modeling confirms that the reduced firing rates in the preferred direction probably underlies the age-related reduction in working memory in monkeys and humans (4). The rate of CUE neuron firing was unaffected by age, which indicates that changes in DELAY neuron function are specific (4). Aging does not diminish the function of all neurons equally.

Wang *et al.* (4) hypothesized that because alpha-2-adrenergic receptors are lost in the prefrontal cortex during aging (2), and alpha-2A–expressing neurons inhibit cyclic adenosine monophosphate (cAMP) signaling, increasing alpha-2A noradrenergic signaling or decreasing cAMP signaling would restore DELAY neuron activity and working memory in old monkeys. cAMP activity is normally inhibited by signaling from alpha-2A noradrenergic receptors and is higher in their absence (**Figure 5.1**) (5). Increased cAMP opens hyperpolarization-activated cyclic nucleotide-gated (HCN) potassium channels and stimulates protein kinase A (PKA), which opens delayed rectifier volt- age-gated potassium channels (KCNQ). When open, these potassium channels weaken persistent neural firing. Increased cAMP signaling also weakens the dendritic spines in layer III, where the self-excitatory working memory neural networks are located.

To test their hypothesis, Wang *et al.* (4) delivered minute amounts of drugs that alter cAMP signaling, alpha-2A activity, or calcium channel opening directly to neurons in area 46 via iontophoresis. Guanfacine, an alpha-2A agonist, and Rp-cAMPS, a cAMP/PKA inhibitor, both significantly increased DELAY neuron firing in old animals. Consistent with the hypothesis, etazolate, a PDE4 inhibitor that increases cAMP, even further reduces DELAY neuron firing in old animals (4). Both HCN channel blocker ZD7288 and KCNQ channel blocker XE991 also increased DELAY neuron firing in old animals. These experiments are consistent with rat experiments showing that guanfacine and Rp-cAMPS enhance working memory in rats[6,7] and support the model in **Figure 5.1**.

Biochemical Mechanisms of Memory Loss in the Aging Brain

To put these results into perspective, Wang *et al.*[4] have provided evidence that specific changes in a subset of cells (alpha-2A receptors) affect cells downstream to cause dysfunction in working memory and that these changes can be overcome by specific modification of the cellular environment by chemical intervention. It is unclear whether age-associated downregulation of alpha-2A noradrenergic receptors is due to molecular damage or unstable epigenetic maintenance of the differentiated state. However, the specificity of the effect argues that key aspects of aging may be due to specific molecular dysfunction, at least in some cases.

More evidence that the memory changes associated with aging have specific biochemical mechanisms is that the loss of ability to recall episodes (episodic memory) with age is independent of working memory in the prefrontal cortex (8). Because episodic memory processing occurs largely in the hippocampus, short-and long-term memory are localized differentially. Of great interest is that compounds that increase cAMP levels and PKA activity by inhibiting cAMP breakdown improve episodic memory in old mice (9,10). Thus, contrary to its effect in the prefrontal cortex, increased cAMP signaling increases memory function in the hippocampus. A possible explanation for this apparent contradictory behavior of cAMP on memory is that increased PKA activity may help create enduring changes in synaptic connections, which is useful for the hippocampal-based long-term memory but counterproductive for short-term working memory (8).

Medical Implications

The use of alpha-2A receptor agonists, such as guanfacine, a Food and Drug Administration (FDA)–approved drug for attention-deficit disorder may prove useful for elderly patients with working memory dysfunction, and a clinical trial is underway to test this hypothesis.

On the basis of work reported in Wang *et al.*,(4) it would be useful to develop drugs to upregulate the levels of alpha-2A receptors in noradrenergic neurons to address the apparent primary defect that underlies the reduction in working memory. Furthermore, it would be quite interesting to determine the mechanism by which the alpha-2A receptors are downregulated with age. Agents that affect cAMP, PKA levels, and potassium channels are not likely to be useful because of their opposing effects on working and long-term memory, unless they can be localized to the appropriate brain substructures.

Conclusions

The use of alpha-2A receptor agonists, such as guanfacine, represents a promising approach to increase working memory in the elderly. The possible effect of guanfacine as a prophylactic to prevent loss of working memory has not been studied, and its ability

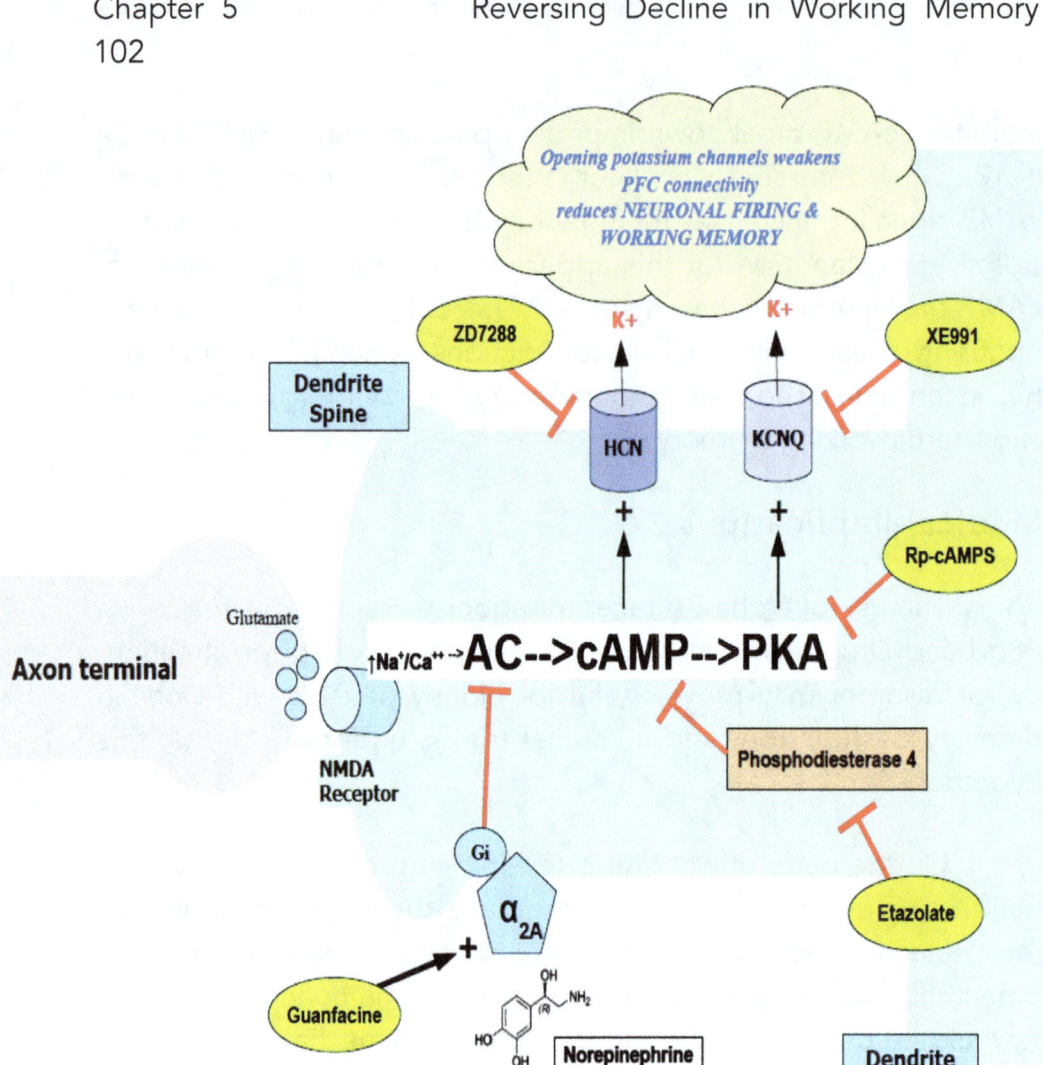

Figure 5.1 **Molecular signaling by cyclic adenosine monophosphate (cAMP) in the spines of prefrontal cortex (PFC) pyramidal dendrite spines regulates network strength and working memory.** Pyramidal cells synapse on spines where cAMP/protein kinase A (PKA) signaling regulates the open state of HCN and KCNQ channels to modulate strength of network connections. Experimental interventions supporting the work of Wang et al.4 are shown (yellow ovals). HCN, Hyperpolarization-activated cyclic nucleotide potassium channel; KCNQ, delayed rectifier voltage-gated potassium channel; AC, adenyl cyclase; PKA, protein kinase A; α2A, adrenergic receptor; NMDA, N-methyl-D-aspartate receptor; Gi, inhibitory G protein.

increase spatial working memory in healthy young humans is unclear because contradictory data have been reported (11,12). This work reinforces a critical insight—critical phenotypes associated with aging may be due to specific molecular changes and not large-scale cell destruction.

References

1. Moore TL, Killiany RJ, Herndon JG, Rosene DL, Moss MB. Executive system dysfunction occurs as early as middle-age in the rhesus monkey. Neurobiol Aging 2006;27:1484–1493.
2. Moore TL, Schettler SP, Killiany RJ, Herndon JG, Luebke JI, Moss MB, Rosene DL. Cognitive impairment in aged rhesus monkeys associated with monoamine receptors in the pre-frontal cortex. Behav Brain Res 2005;160:208–221.
3. Hara Y, Rapp PR, Morrison JH. Neuronal and morphological bases of cognitive decline in aged rhesus monkeys. Age (Dordr) 2011; Jun 28. [Epub ahead of print, PMID:21710198]
4. Wang M, Gamo NJ, Yang Y, Jin LE, Wang X-J, Laubach M, Mazer JA, Lee D, Arnsten AFT. Neuronal basis of age-related working memory decline. Nature 2011;476:210–213.
5. Downs JL, Dunn MR, Borok E, Shanabrough M, Horvath TL, Kohama SG, Urbanski HF. Orexin neuronal changes in the locus coeruleus of the aging rhesus macaque. Neurobiol Aging 2007;28:1286–1295.
6. Ramos BP, Stark D, Verduzco L, van Dyck CH, Arnsten AFT. Alpha2a-adrenoceptor stimulation improves prefrontal cortical regulation of behavior via inhibition of camp signaling in aging animals. Learn Mem 2006;13:770–776.
7. Ramos BP, Birnbaum SG, Lindenmayer I, Newton SS, Duman RS, Arnsten AFT. Dysregulation of protein kinase a signaling in the aged prefrontal cortex: New strategy for treating age-related cognitive decline. Neuron 2003;40:835–845.
8. Baxter MG. Age-related memory impairment. Is the cure worse than the disease? Neuron 2003;40:669–670.
9. Barad M, Bourtchouladze R, Winder DG, Golan H, Kandel E. Rolipram, a type IV-specific phosphodiesterase inhibitor, facilitates the establishment of long-lasting long-term potentiation and improves memory. Proc Natl Acad Sci USA 1998;95:15020–15025.
10. Bach ME, Barad M, Son H, Zhuo M, Lu Y-F, Shih R, Mansuy I, Hawkins RD, Kandel ER. Age-related defects in spatial memory are correlated with defects in the late phase of hippocampal long-term potentiation in vitro and are attenuated by drugs that enhance the cAMP signaling pathway. Proc Natl Acad Sci USA

1999;96:5280–5285.
11. Jakala P, Riekkinen M, SirviU J, Koivisto E, Kejonen K, Vanhanen M, Riekkinen P Jr. Guanfacine, but not clonidine, improves planning and working memory performance in humans. Neuropsychopharmacology 1999;20:460–470.
12. Muller U, Clark L, Lam ML, Moore RM, Murphy CL, Richmond NK, Sandhu RS, Wilkins IA, Menon DK, Sahakian BJ, Robbins TW. Lack of effects of guanfacine on executive and memory functions in healthy male volunteers. Psychopharmacology (Berl) 2005;182:205–213.

Chapter 6

Dissecting mammalian target of rapamycin to promote longevity

Treatment with rapamycin, an inhibitor of mammalian target of rapamycin complex 1 (mTORC1) can increase mammalian life span. However, extended treatment with rapamycin results in increased hepatic gluconeogenesis concomitant with glucose and insulin insensitivity through inhibition of mTOR complex 2 (C2). Genetic studies show that increased life span associated with mTORC1 inhibition can be at least partially decoupled from increased gluconeogenesis associated with mTORC2 inhibition. Adenosine monophosphate kinase (AMPK) agonists such as metformin, which inhibits gluconeogenesis by downregulating expression of glucose-6-phosphatase and phosphoenolpyruvate carboxykinase, might be expected to block the glucose dysmetabolism mediated by rapamycin. The search for inhibitors of the mTORC1 component Raptor may prove a productive approach to create a better mTOR inhibitor.

Inhibition of Target of Rapamycin Extends Life Span and Health Span

Inhibition of the target of rapamycin (TOR) pathway extends health span and life span in a variety of organisms, including *Saccharomyces cerevisiae*, *Caenorhabditis elegans*, *Drosophila melanogaster*, and *Mus musculus* (1,2). Rapamycin, an anticancer and immunosuppressant drug, inhibits mammalian TOR (mTOR), a serine-threonine protein kinase that plays a central role in energy/nutrient/redox sensing, protein synthesis, and autophagy. Rapamycin is one of a small number of drugs that has been shown to increase longevity in mammals, even when treatment is restricted to adults (3,4). The mechanisms that underlie this intriguing result are under intense investigation.

mTOR is found in two signaling complexes, mTOR complex 1 (C1) and mTOR complex 2 (C2) (see **Figure 6.1**). mTORC1 acts as an energy, nutrient, growth factor, mitogen, stress, and redox sensor that controls protein synthesis and autophagy in a reciprocal fashion (5). For example, mTORC1 is activated by increased energy and nutrients to positively regulate S6 kinase (S6K) and eukaryotic translation initiation factor 4e-binding protein (eIF4e-BP) (4E-BP1) to initiate and augment translation. Autophagosomal regulators Unc51-like kinase 1 (ULK1) and ATG13 are inhibited by mTORC1 phosphorylation. Although less well characterized, mTORC2 responds to growth factors to regulate cell polarization via the actin cytoskeleton and cell survival and metabolism via phosphorylation of prosurvival Akt, protein kinase C (PKC), and glucocorticoid-regulated kinase (SGK) (6). Akt promotes survival by phosphorylating the proapoptotic transcription factors FOXO1 and FOXO3 to inhibit their translocation to the nucleus. Activation of mTORC2 can also contribute to increased activity of mTORC1 via Akt activation of TSC2 and Rheb4 (see **Figure 6.1**). The immunosuppressive macrolide rapamycin, which binds the immunophilin FK506-binding protein (FKBP12), specifically inhibits mTORC1, but not mTORC2, by interfering with the interaction of mTOR with the mTORC1-specific Raptor protein (5,6). Unfortunately, this picture is muddied in that chronic high doses of rapamycin can also inhibit mTORC2 in some cell types (see below) (7,8). Inhibition of mTORC2 has not been thought to play a role in rapamycin-induced increased longevity, with the exception of a report that *C. elegans* with reduced amounts of Rictor and therefore reduced mTORC2 activity live longer, apparently due to altered intestinal function (9). However, it remains unclear whether this longevity mechanism is evolutionarily conserved.

Importantly, prolonged treatment with rapamycin has been

reported to induce glucose intolerance and increase insulin resistance in mice, rats, and possibly humans (10–14). These effects on glucose homeostasis are the opposite of what would be expected for a drug that extended life span. Typically, augmented insulin sensitivity and reduced phosphoinositide 3 (PI3) kinase activity are associated with increased longevity resulting from dietary restriction or prolongevity genetic mutations in invertebrates (15).

In an important study, Lamming *et al.* (16) show that rapamycin treatment of mice at the same dose that extends life span augments liver-specific expression of gluconeogenesis enzymes phosphoenolpyruvate carboxykinase (PEPCK) and glucose 6-phosphatase (G6Pase), resulting in glucose intolerance and hepatic insulin resistance. Rapamycin-induced glucose intolerance was shown to be independent of mTORC1, the conventional target of rapamycin, using a mouse strain that carried a liver-specific conditional allele of mTORC1 component Raptor (**Figure 6.1**). Because mTORC2 was known to be inhibited by rapamycin in some cell types (8), mTORC2 became the prime suspect for causing the altered glucose homeostasis. Consistent with this hypothesis, phosphorylation of the mTORC2 preferred substrate PKC-a S657 and SGK substrate NDRG1 were attenuated in liver, muscle, and white adipose tissue. Furthermore, immunoprecipitation experiments indicated that rapamycin decreased the association of mTOR with both mTORC1 and mTORC2, supporting the contention that rapamycin was indeed effecting mTORC2. Genetic disruption in mouse strains carrying a liver-specific conditional allele of Rictor (found only in mTORC2), resulted in pronounced glucose intolerance, but only in mild insulin intolerance, which the authors speculate results from the presence of functional Rictor and mTORC2 in nonhepatic tissue. It was already known that disruption of Rictor in adipose tissue caused weight gain, hyperinsulinemia, and insulin resistance (17,18) In the mouse strain carrying a conditional allele of Rictor that could be deleted in all tissue types, glucose homeostasis was altered in the same direction as with prolonged treatment with rapamycin. In addition, additional rapamycin had no further effect on these mice (16), consistent with the hypothesis that rapamycin-mediated inhibition of mTORC2 causes glucose dysregulation.

Rapamycin-Mediated Life Span Extension Can Be Decoupled from Rapamycin-Mediated Glucose Dysregulation

The naive expectation from numerous genetic and dietary restriction studies on longevity in invertebrates is that insulin sensitivity and inhibition of insulin signaling would extend life span. Therefore, the authors hypothesized that rapamycin-mediated glucose dysregulation was likely not involved in life span extension. To test this hypothesis, they bred mice that carried only single copies (haploinsufficiency) of mtor, Raptor, or mlst8, or the double mutations mtor+/-, Raptor +/- and mtor +/-, mlst8 +/- in an effort to find a differential effect on mTORC1 and mTORC2 signaling that correlated with longevity. They could not perform full genetic knockouts of mtor, Raptor, Rictor, or mlst8 that might have resulted in stronger phenotypes, because these phenotypes are embryonic lethal. It is unclear why they did not also breed mice with a single copy of Rictor. Interestingly, mice carrying a single allele of Raptor either alone or in combination with a single copy of mTOR exhibited a normal life span—a result not consistent with rapamycin acting through mTORC1. However, the authors speculate that the approximate 50% inhibition of mTORC1 is not sufficient to extend life span and that rapamycin results in greater inhibition of mTORC1 than observed here (16).

Only female mtor+/-, mlst8+/- mice were long lived (14.4% increase in mean life span). In these mice, hepatic mTORC1 signaling was decreased by about 50%, but mTORC2 signaling was

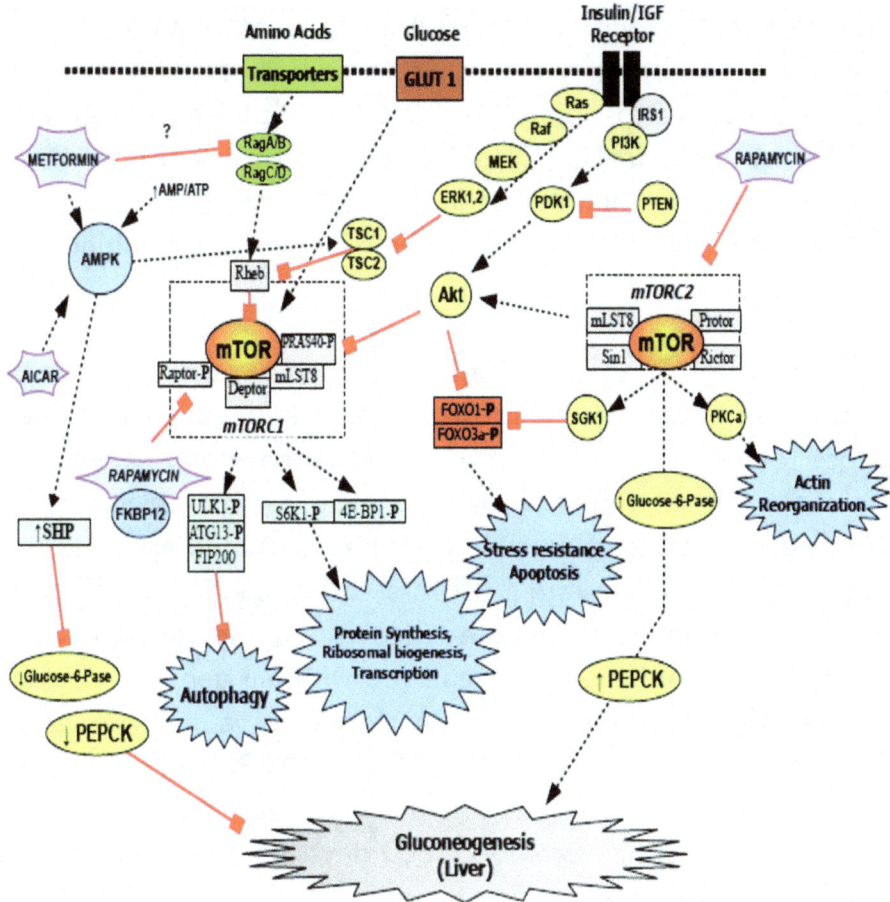

Figure 6.1 Simplified schematic of mammalian target of rapamycin (mTOR) signaling pathways and their interaction with rapamycin. mTOR complex 1 (C1) promotes messenger RNA (mRNA) translation and inhibits autophagy by integrating growth factor signals like insulin/insulin-like growth factor (IGF), amino acid nutrient signals, and adenosine triphosphate (ATP) levels via 5′-adenosine monophosphate-activated protein kinase (AMPK). mTORC1 affects many homeostatic processes beyond cell growth and autophagy, including neural memory formation/consolidation and immunity. Growth factors activate mTORC2 to activate Akt and Sgk1, which promote survival and inhibit forkhead transcription factors FOXO1 and FOXO3. The forkhead transcription factors promote antioxidant enzyme expression and cell death. Rapamycin can inhibit mTORC2 to stimulate gluconeogenesis by upregulating glucose 6-phosphatase (G6Pase) and phosphoenolpyruvate carboxykinase (PEPCK). AMPK agonist metformin inhibits gluconeogenesis by downregulating transcription of G6Pase and PEPCK. At the same time, metformin also can inhibit mTORC1.

unaffected, even though both mTOR complexes might have a priori been thought to be affected by single copies of mTOR and mlst8 and given that in Drosophila the mlst8 homolog lst8 is required for mTORC2, but not mTORC1 signaling. The explanation may be related to the observation that the ratio of Raptor to mTOR declined more than the ratio of Rictor to mTOR in these mice, resulting in less mTORC1 signaling (16).

Glucose tolerance was not affected in any of the strains with the exception that male Raptor +/- mice actually had increased glucose tolerance. The normal glucose tolerance and increased life span of the female mtor +/- mlst8 +/- mice, and the unchanged life span despite increased glucose tolerance in male Raptor +/- mice suggests that rapamycin's effects on longevity and glucose homeostasis can be uncoupled. However, the result is not as satisfying as might have been achieved with full genetic knockouts. Perhaps the experiment can be reinforced by a study in which a mouse strain carrying a ubiquitous conditional Rictor and/or Raptor knockout is treated with rapamycin just after inducing genetic deletion when the mice are about 600 days old, the same age used in Harrison *et al.*'s life-extending experiments (3).

Rapamycin as an Antiaging Therapeutic Revisited

Recently, we reviewed the advantages and disadvantages of rapamycin as a potential anti-aging therapeutic (19). The risk of infections and cancer from immune suppression are probably offset by rapamycin's significant anticancer and antiviral activities (20–23). Some of the most serious reported potential adverse side effects, such as hypertriglyceridemia and hypercholesterolemia, may be due to mTORC2 inhibition. However, stimulation of gluconeogenesis and the promotion of insulin resistance resulting from mTORC2 inhibition remain perhaps the most serious impediments to using rapamycin as an anti-aging therapeutic. There are two possible routes to overcome this problem:

(1) Stimulate mTORC2 or downstream pathways involved in glucose regulation sufficiently to overcome rapamycin's inhibitory effect or

(1) develop analogs of rapamycin (rapalogs) or new drugs that specifically target only mTORC1.

Perhaps the most readily available means to achieve the first goal would be to augment rapamycin therapy with metformin, an AMP kinase (AMPK) agonist that inhibits gluconeogenesis. Metformin stimulates AMPK, which in turn increases expression of SHP, a transcriptional repressor of genes encoding glucose-6-phosphatase and phospho-enolpyruvate carboxykinase, key enzymes of gluconeogenesis (**Figure 6.1**) (24). Overexpression of SHP will probably have a dominant effect on gluconeogenesis, overcoming the pro-gluconeogenesis effects mediated by mTORC2 activation, but this hypothesis will need to be established experimentally and then confirmed clinically. Interestingly, metformin co-treatment will also contribute to mTORC1 inhibition: MPK stimulates TSC1/2 to convert Rheb to its inactive guanosine diphosphate (GDP) form, which in turn inhibits mTORC1 (**Figure 6.1**) (25). Fortunately, a Phase I trial exploring the combination of rapamycin with metformin in treatment of advanced cancers is currently recruiting (NCT01529593, clinicaltrials.gov). The AMPK activator 5-aminoimidazole-4-carboxamide-1-b-D-ribofurano- side (AICAR) may have similar activity to metformin (**Figure 6.1**).

Low doses of resveratrol have been reported to stimulate mTORC2 activity (26) and may be useful in combination with rapamycin. However, independent confirmation of these results is necessary, given the controversy over the integrity of reports from this group (27). As for the second goal of developing mTORC1-specific drugs: Currently available rapalogs and related compounds, developed for anticancer therapy are unlikely to be useful in their current form because they have been designed to inhibit both mTORC1 and mTORC2, e.g., mTOR kinase inhibitors that block catalysis (28). Drug screens that target the mTORC1- specific component Raptor would be a productive way to find mTORC1-specific inhibitors.

Conclusions

Rapamycin has been shown to inactivate mTORC1 to promote longevity and mTORC2, at least in the liver, to alter glucose homeostasis. The longevity effect is separable from the effect on glucose homeostasis. The development of specific inhibitors of mTORC1 or combinations of rapamycin with drugs that suppress gluconeogenesis and insulin resistance may lead to the first drug-based rational antiaging therapeutic regimen.

References

1. Bjedov I, Partridge L. A longer and healthier life with TOR down-regulation: Genetics and drugs. Biochem Soc Trans 2011;39:460–465.
2. Hands SL, Proud CG, Wyttenbach A. mTOR's role in ageing: Protein synthesis or autophagy? Aging (Albany NY) 2009;1:586–597.
3. Harrison DE, Strong R, Sharp ZD, Nelson JF, Astle CM, Flurkey K, Nadon NL, Wilkinson JE, Frenkel K, Carter CS, Pahor M, Javors MA, Fernandez E, Miller RA. Rapamycin fed late in life extends lifespan in genetically heterogeneous mice. Nature 2009;460:392–395.
4. Miller RA, Harrison DE, Astle CM, Baur JA, Boyd AR, de Cabo R, Fernandez E, Flurkey K, Javors MA, Nelson JF, Orihuela CJ, Pletcher S, Sharp ZD, Sinclair D, Starnes JW, Wilkinson JE, Nadon NL, Strong R. Rapamycin, but not resveratrol or simvastatin, extends life span of genetically heterogeneous mice. J Gerontol A Biol Sci Med Sci 2011;66: 191–201.
5. Laplante M, Sabatini DM. Mtor signaling at a glance. J Cell Sci 2009;122:3589–3594.
6. Zoncu R, Efeyan A, Sabatini DM. mTOR: From growth signal integration to cancer, diabetes and ageing. Nat Rev Mol Cell Biol 2011;12:21–35.
7. Foster DA, Toschi A. Targeting mTOR with rapamycin: One dose does not fit all. Cell Cycle 2009;8:1026–1029.
8. Sarbassov DD, Ali SM, Sengupta S, Sheen J-H, Hsu PP, Bagley AF, Markhard AL, Sabatini DM. Prolonged rapamycin treatment inhibits mTORC2 assembly and akt/pkb. Mol Cell 2006;22:159–168.

9. Soukas AA, Kane EA, Carr CE, Melo JA, Ruvkun G. Rictor/ TORC2 regulates fat metabolism, feeding, growth, and life span in Caenorhabditis elegans. Genes Dev 2009;23:496–511.
10. Cunningham JT, Rodgers JT, Arlow DH, Vazquez F, Mootha VK, Puigserver P. mTOR controls mitochondrial oxidative function through a yy1-pgc-1alpha transcriptional complex. Nature 2007;450:736–740.
11. Fraenkel M, Ketzinel-Gilad M, Ariav Y, Pappo O, Karaca M, Castel J, Berthault M-F, Magnan C, Cerasi E, Kaiser N, Leibowitz G. mTOR inhibition by rapamycin prevents beta-cell adaptation to hyperglycemia and exacerbates the metabolic state in type 2 diabetes. Diabetes 2008;57:945–957.
12. Houde VP, Brule S, Festuccia WT, Blanchard P-G, Bellmann K, Deshaies Y, Marette A. Chronic rapamycin treatment causes glucose intolerance and hyperlipidemia by upregulating hepatic gluconeogenesis and impairing lipid deposition in adipose tissue. Diabetes 2010;59:1338–1348.
13. Johnston O, Rose CL, Webster AC, Gill JS. Sirolimus is asociated with new-onset diabetes in kidney transplant recipients. J Am Soc Nephrol 2008;19:1411–1418.
14. Teutonico A, Schena PF, Di Paolo S. Glucose metabolism in renal transplant recipients: Effect of calcineurin inhibitor withdrawal and conversion to sirolimus. J Am Soc Nephrol 2005;16:3128–3135.
15. Kenyon CJ. The genetics of ageing. Nature 2010;464:504–512.
16. Lamming DW, Ye L, Katajisto P, Goncalves MD, Saitoh M, Stevens DM, Davis JG, Salmon AB, Richardson A, Ahima RS, Guertin DA, Sabatini DM, Baur JA. Rapamycin-induced insulin resistance is mediated by mTORC2 loss and un-coupled from longevity. Science 2012;335:1638–1643.
17. Kumar A, Lawrence JC Jr, Jung DY, Ko HJ, Keller SR, Kim JK, Magnuson MA, Harris TE. Fat cell-specific ablation of Rictor in mice impairs insulin-regulated fat cell and whole- body glucose and lipid metabolism. Diabetes 2010;59:1397–1406.
18. Cybulski N, Polak P, Auwerx J, Ru egg MA, Hall MN. mTOR complex 2 in adipose tissue negatively controls whole-body growth. Proc Natl Acad Sci USA. 2009;106:9902–9907.
19. Mendelsohn AR, Larrick JW. Rapamycin as an antiaging therapeutic?: Targeting mammalian target of rapamycin to treat hutchinson-gilford progeria and neurodegenerative diseases. Rejuvenation Res 2011;14:437–441.
20. Witzig TE, Reeder CB, LaPlant BR, Gupta M, Johnston PB, Micallef IN, Porrata LF, Ansell SM, Colgan JP, Jacobsen ED, Ghobrial IM, Habermann TM. A phase II trial of the oral mTOR inhibitor everolimus in relapsed aggressive lymphoma. Leukemia 2011;25:341–347.
21. Leblanc KG Jr, Hughes MP, Sheehan DJ. The role of sirolimus in the prevention of cutaneous squamous cell carcinoma in organ transplant recipients. Dermatol Surg 2011;37: 744–749.

22. Dancey J. Mtor signaling and drug development in cancer. Nat Rev Clin Oncol 2010;7:209–219.
23. Cervera C, Fernandez-Ruiz M, Valledor A, Linares L, Anton A, Angeles Marcos M, Sanclemente G, Hoyo I, Cofan F, Ricart MJ, Perez-Villa F, Navasa M, Pumarola T, Moreno A. Epidemiology and risk factors for late infection in solid organ transplant recipients. Transpl Infect Dis 2011;13:598– 607.
24.. Kim YD, Park K-G, Lee Y-S, Park Y-Y, Kim D-K, Nedumaran B, Jang WG, Cho W-J, Ha J, Lee I-K, Lee C-H, Choi H-S. Metformin inhibits hepatic gluconeogenesis through AMP-activated protein kinase-dependent regulation of the orphan nuclear receptor SHP. Diabetes 2008;57:306–314.
25. Kalender A, Selvaraj A, Kim SY, Gulati P, Brule S, Viollet B, Kemp BE, Bardeesy N, Dennis P, Schlager JJ, Marette A, Kozma SC, Thomas G. Metformin, independent of AMPK, inhibits mTORC1 in a rag GTPase-dependent manner. Cell Metab 2010;11:390–401.
26. Gurusamy N, Lekli I, Mukherjee S, Ray D, Ahsan MK, Gherghiceanu M, Popescu LM, Das DK. Cardioprotection by resveratrol: A novel mechanism via autophagy involving the mTORC2 pathway. Cardiovasc Res 2010;86:103–112.
27. Roehr B. Cardiovascular researcher fabricated data in studies of red wine. BMJ 2012;344:e406.
28. Wang X, Sun S-Y. Enhancing mTOR-targeted cancer therapy. Expert Opin Ther Targets 2009;13:1193–1203.

Chapter 7

Dietary restriction: critical co-factors to separate health span from life span benefits

Dietary restriction (DR), typically a 20%–40% reduction in ad libitum or "normal" nutritional energy intake, has been reported to extend life span in diverse organisms, including yeast, nematodes, spiders, fruit flies, mice, rats, and rhesus monkeys. The magnitude of the life span enhancement appears to diminish with increasing organismal complexity. However, the extent of life span extension has been notoriously inconsistent, especially in mammals. Recently, Mattison *et al.* reported that DR does not extend life span in rhesus monkeys (1) in contrast to earlier work of Colman *et al.* (2). Examination of these papers identifies multiple potential confounding factors. Among these are the varied genetic backgrounds and composition of the "normal" and DR diets. In monkeys, the correlation of DR with increased health span is stronger than that seen with life span and indeed may be separable. Recent mechanistic studies in Drosophila (3) implicate non-genetic co-factors such as level of physical activity and muscular fatty acid metabolism in the benefits of DR. These results should be followed up in mammals. Perhaps levels of physical activity among the cohorts of rhesus monkeys contribute to inconsistent DR effects. To understand the maximum potential benefits from DR requires differentiating fundamental effects on aging at the cellular and molecular levels from suppression of age-associated diseases, such as cancer. To that end, it is important that investigators carefully evaluate the effects of DR on biomarkers of molecular aging, such as mutation rate and epigenomic alterations. Several short-term studies show that humans may benefit from DR in as little as 6 months, by achieving lowered fasting insulin levels and improved cardiovascular health. Optimized health span engineering will require a much deeper understanding of DR.

Introduction

Caloric or dietary restriction (DR), typically a 20%–40% reduction in *ad libitum* or "normal" nutritional energy intake, has been associated with increased life span in diverse organisms including yeast, nematodes, spiders, fruit flies, fish, mice, rats, and rhesus monkeys (4). <u>Note</u> *that we use the terms DR and calorie restriction [CR] synonymously.* The literature sometimes refers to CR to mean decreased caloric intake and DR to mean restriction of one of more components of intake.) First observed in laboratory rats as long ago as 1935 (5), the incredible diversity of organisms that respond to DR with increased life span has suggested that DR fundamentally slows the rate of aging, perhaps by nutritional stress-induced stimulation of pathways involved in survival. However, besides the obvious problems in consistently defining what constitutes aging beyond longevity, critics have pointed out that "normal" diet is a nebulous concept and that DR may merely define an optimal diet. Moreover, debate continues over whether DR actually slows the rate of aging versus only delaying the onset of age-associated diseases, at least in mammals. Additional relevant questions concern whether the effects of DR are truly universal and to what extent should results from invertebrates or even short-lived mammals correctly inform human health, given the large physiological differences.

The comparative maximum benefit of DR to life span appears to decrease with increasing organismal complexity. DR can increase life span of yeast three-fold, nematodes two- to three-fold, Drosophila two-fold, and mice 30%–50% (6). Maximal benefit is achieved when DR is begun early in life, and the later a DR regimen is initiated, the less the benefit for any particular species.

Mutations and drugs that confer longevity benefits have helped identify the key targets of DR. Especially in invertebrates, such studies have given credence to DR's possible universality and provide the strongest evidence that DR actually affects the rate of aging. Evolutionarily conserved molecular mechanisms for dietary restriction result in inhibition of nutrient-sensing pathways and alter organismal and cellular metabolism in yeast, worms, flies and mammals (**Figure 7.1**). Specifically, inhibition of the target of rapamycin (TOR)/ribosomal S6 protein kinase (S6K) pathways by genetic knockout or by the TOR/mammalian (m)TOR inhibitor rapamycin results in increased life span in yeast, worms, flies, and mice. In worms and flies, rapamycin-mediated life span increase requires the conserved "metabolic brake" eukaryotic initiation factor 4E binding protein (4E-BP), autophagy (which increases with decreased mTOR/S6K activity), and reduced (S6K) activity. Low levels of nutrients associated with DR similarly inhibit the TOR pathway. Interestingly, rapamycin treatment of adult mice results in a 10% increase in life span in older adult mice,7 which may even exceed the benefit of DR in that model. In multi-cellular animals, DR inhibits insulin/insulin-like growth factor-1 (IGF-1) (IIS) pathways, which leads to increased increased life span via inhibition of phosphoinositide 3-kinase (PI3K) and Akt, and stimulation of master transcription factor DAF-16 (worms)/Foxo (flies, mammals?), which in turn activates multiple protective downstream pathways.

Ultimately, the combination of IIS and TOR/S6K inhibition results in increased levels of enzymes that protect cells from oxidative damage such as superoxide dismutase (SOD) and catalase (except flies), proteins that stabilize protein conformation such as heat shock proteins/chaperones, autophagy, endoplasmic reticulum (ER) stress response, xenobiotic metabolism, and reduced translation and fat accumulation (6). Reduced levels of amino acids are thought to predominately inhibit the mTOR pathway, whereas glucose restriction targets the IIS pathway (**Figure 7.1**).

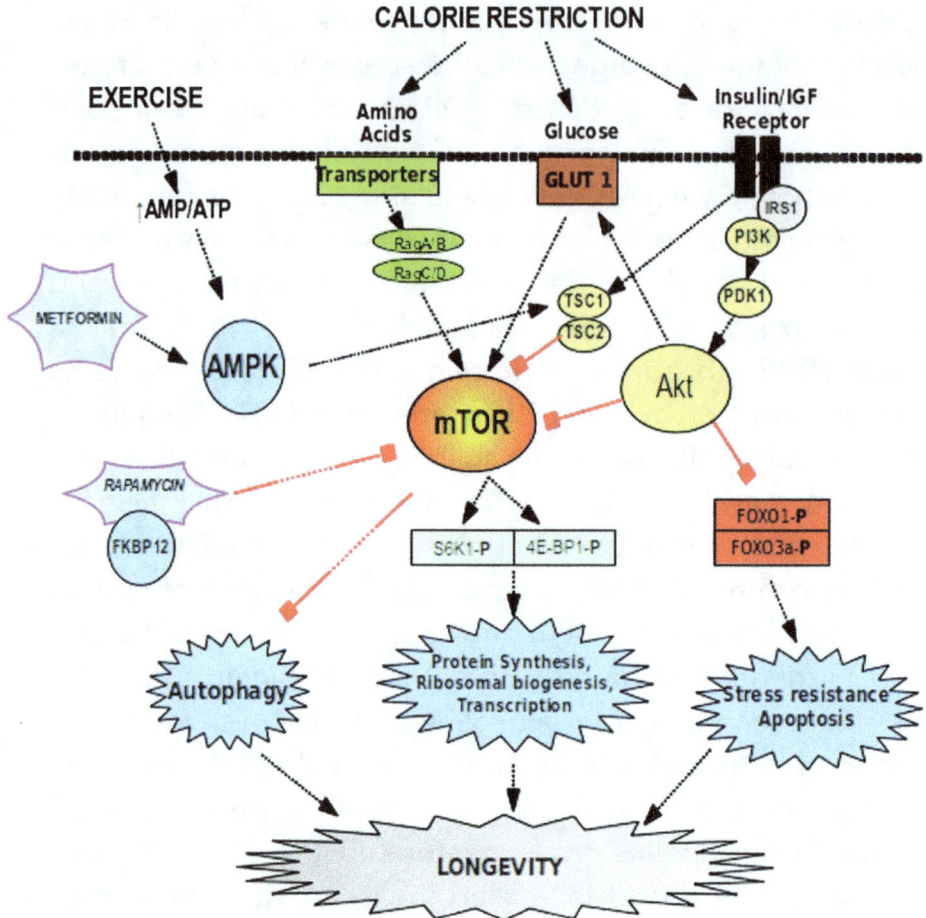

Figure 7.1 Evolutionarily conserved molecular mechanisms for dietary restriction (DR). Inhibition of nutrient-sensing pathways alter organismal and cellular metabolism in yeast, worms, flies, and mammals. DR down-regulates the target of rapamycin (TOR)/mammalian (m)TOR pathway and the insulin/insulin-like growth factor-I (IGF-1) (IIS) pathways leading to increased levels of superoxide dismutase (SOD), catalase (except files), heat shock proteins/chaperones, autophagy, endoplasmic reticulum (ER) stress response, and reduced translation and fat accumulation. Reduced sugar/insulin/IGF-1 eventually leads to increased levels of Daf16 (worms)/Foxo (mammals, flies), which acts on multiple downstream pathways to effect increased longevity. Reduced levels of amino acids, especially methionine, are thought to predominately inhibit the mTOR pathway, whereas glucose restriction not only targets the Daf16/IGF-1 pathway, but also activates master nutrient sensor adenosine monophosphate (AMP) kinase. Rapamycin, an inhibitor of TOR, extends life span in adult mice, at least as well as DR does. ATP, Adenosine triphophate; PI3K, phosphoinositide 3-kinase; AMPK, AMP-activated protein kinase.

However, in humans, unlike other studied multicellular animals including rodents, DR by itself does not lower the IIS growth factor IGF-1; instead reducing protein intake lowers IGF-1,8 suggesting that humans may have altered mTOR/IIS regulation.

Increasing organismal complexity not only diminishes maximum benefit of DR on life span, but perhaps also decreases the consistency of the life span enhancement, which has created controversy as to the value of DR to extending human life span. However, even in invertebrates, DR has not always resulted in life span enhancement. Reported exceptions include *Drosophila melanogaster* (which more often than not, actually do respond to DR), the nematode *Caenorhabditis remanei* (unlike its cousin *C. elegans*), medflies, butterflies, the housefly, and the spider *Latrodectus hasselti*. In mammals, specifically rodents, the results are even more problematic. A meta-review of the literature by Swindell reveals that DR works more consistently in rats (median of experiments support a 14%–45% increase) than in mice (median of experiments support a 8% decrease in life span to a 20% increase) (9). Especially problematic are mouse studies in which life span decreases or remains unaffected and those in which many animals die from non-aging effects. It is not reassuring that non-inbred mouse strains newly derived from wild animals tend to not respond to DR.

Unfortunately, technical differences in DR protocols, including diet composition and varied genetic backgrounds of the animals may account for much of the inconsistency, which makes interpretation of the data more difficult (9). Interestingly, even in cases where life span enhancement is not seen, often health span benefits are seen, including reduction of age-associated pathologies such as cancer, heart disease, and kidney disease, suggesting DR may have medical value independent of increasing life span.

Effect of DR on Non-Human Primates

If the maximum DR effect on life span does diminish with organismal complexity, perhaps it is because more complex organisms already make better use of the stress signaling pathways to maintain homeostasis. More complex animals have a greater need to coordinate responses efficiently among a large number of cells when subjected to stresses such as starvation, to reduce cell division/reproduction, and conserve energy. To assess the relevance of DR to humans requires study of animals more closely related to humans, such as other primates. Life span studies on humans are not practical, but two studies on rhesus monkeys have been ongoing since the 1980s at the National Institute of Aging (NIA) and the Wisconsin National Primate Research Center (WNPRC). Initial results from the WNPRC, reported in 2009, suggested that young animals (7–14 years old) undergoing DR had a trend toward a longer life span. This trend becomes statistically significant if animals that died from nonaging causes are not considered. Most importantly, they observed that DR significantly reduced the incidence of age-associated diseases such as diabetes, cancer, cardiovascular disease, and brain atrophy (2). In an important recent study, Mattison *et al.* at the NIA report that contrary to the WNPRC study, DR had no effect on the survival of rhesus monkeys, regardless of whether DR was initiated in young monkeys (1–14 years old) or older animals (16–23 years old). Although 50% of young animals remain alive, statistical projection suggests that less than 0(1)% of the DR animals will outlive the controls. Some health span–related parameters did improve: DR-treated older animals had lower triglycerides and glucose levels and DR-treated males had significantly lower cholesterol levels than untreated males. However, cholesterol levels in DR-treated females were unaffected, and in young DR animals, glucose levels were unaffected, whereas triglycerides were slightly lower only in males (1).

To explain the contradictory results between the NIA and

WNPRC results, differences in diet have been invoked: The untreated WNPRC animals were allowed to feed ad libitum and as a consequence weighed 10% more than the NIA control animals, suggesting that the control animals in the NIA study may already be subject to a weak DR effect. Interestingly, the weight of rhesus monkey in both studies exceeds that of wild rhesus monkeys, raising again the problem of what constitutes a "normal" diet. Also, the NIA diet may have been healthier: Sucrose made up 28.5% of the WNPRC animal diet, but only 3.9% of the NIA diet, which may explain an increase in diabetes incidence (40% in the WNPRC vs. 12.5% in the NIA) (10). The NIA diet also included more plant-derived micro-nutrients and phytochemicals than the WNPRC diet. It is also unclear if differences in the physical activity levels of the animals in the two studies may have played a role (see below). However, it is clear that DR results in at least modest increased health span in these non-human primate studies.

Physical Activity Levels and Fatty Acid Metabolism in Muscle May Play Critical Roles in DR

Katewa *et al.* show that DR induces an increase in fatty acid synthesis and breakdown in *D. melanogaster*. Inhibition of fatty acid synthesis or oxidation genes inhibited DR-mediated life span extension. Furthermore, this increase in fatty acid metabolic rate correlated with an increase in physical activity that was at least partially required for DR to extend longevity in flies. Increased fatty acid metabolism was monitored by feeding the flies with 14C-labeled glucose for 24 hr. Triglyceride synthesis rates increased 2.8-fold in DR animals, whereas chasing with unlabeled food for 60 hr resulted in a 63% decrease in triglycerides in DR animals. Triglyceride synthesis was unaffected in control animals. Together these results demonstrate that both lipogenesis and lipolysis are increased during DR. Using a strain that expresses an inducible RNA interference (RNAi) that targets acetyl coenzyme A carboxylase (dACC), DR-mediated life span increase was reduced (113% increase in controls to 52% in dACC-inhibited females and from a 22% increase to 5% in dACC-inhibited males). Knockdown of dACC also reduced starvation and cold stress resistance in DR treated flies (3).

Using genome-wide transcription analysis, Katewa *et al.* observed that dACC inhibition reversed DR-mediated changes in a cluster of genes involved in muscle structure and function. To confirm that the effect was muscle specific, Katewa repeated experiments using RNAi targeting dACC in the fat body, neurons, and muscle using inducible promoters specific for each tissue type. Only muscle-specific dACC-targeted RNAi reduced DR-mediated life span enhancement. To dissect this effect, Katewa and colleagues used RNAi knockdown to target CG4389 and CG7834, which encode for mitochondrial long-chain 3-hydroxyacyl coenzyme A dehydrogenase and electron transport flavoprotein b-subunit proteins, respectively. Similar to inhibition of dACC, muscle-specific knockdown of CG4389 or CG7834 significantly reduced the DR-dependent life span extension, demonstrating that increased fatty acid metabolism in muscle is required for DR life span increase. Interestingly, treatment with muscle-specific RNAi targeting dACC, CG7834, or CG4389 also resulted in a significant reduction in DR-associated physical movement (3). DR is reported to enhance spontaneous movement-related activity in diverse species, including flies, rodents, and primates, probably due to selection for increased foraging under conditions of nutritional scarcity. Katewa *et al.* confirmed that their DR flies had higher levels of spontaneous activity, but inhibition of dACC lowered spontaneous activity in DR-treated flies at all ages, showing dACC is required for the increased physical activity. To test whether spontaneous activity is necessary for DR, wing defects were induced genetically by expressing the cell death–inducing gene reaper in wings. In flies with defective wings, DR only induced a very modest 14% life span extension versus 61% for appropriate controls. As an alternate strategy, when flies wings were mechanically clipped, DR-mediated life span was modestly reduced from 97% to 33%. Although the effects of limiting physical activity are less than those seen with RNAi knockdown of dACC, increased physical activity is at least partially required for DR-mediated life span increase.

Finally, consistent with an important role for fatty acid metabolism and physical activity in DR, over-expression of AKH, a glucagon ortholog that maintains glucose and triglyceride homeostasis in flies and mammals, increased fatty acid metabolism, spontaneous physical activity, and life span by 33% in ad libitum–fed flies. As expected, over-expression of AKH did not further increase life span in DR flies (3). These results suggest that modulation of fatty acid metabolism and conditions that allow or limit physical activity may prove important to interpreting DR experiments in other systems, such as the rhesus monkey studies. There is a cautionary note here. In Katewa's study, DR was implemented by reducing the total amount of food available to the flies. When DR in flies involves reducing calories by only lowering sugar, fatty acid metabolism is probably unaffected, which may eliminate a physical activity effect on DR (11).

Medical Implications

Human data

First, it should be clear that invertebrate, rodent, and perhaps even rhesus monkey DR results may not translate well to humans. There are two types of studies that evaluate the potential impact of DR on humans. The first connects genetic mutations in the DR pathways to longevity and health span. Ecuadorians with growth hormone receptor (GHR) mutations that produce severe GHR and IGF-1 deficiencies do not seem to live longer, but are remarkably free of diabetes and cancer and have reduced incidence of stroke (12) Interestingly, insulin and mTOR expression are also reduced (**Figure 7.1**). Alcohol abuse and accidents leading to premature death appear to limit any definitive conclusions regarding the effects this mutation may have on longevity, although it is curious that no exemplar of increased maximum longevity emerged (12).

Although life span studies on humans may be impractical, shorter-term DR human health span studies have been reported, although many are small studies that require replication. The best-controlled human study is the NIA's CALERIE research program. The Phase I CALERIE clinical trial determined the short-term (6 or 12 months) effects of 20% or 25% CR in non-obese humans. DR subjects had decreased whole body and visceral fat, reduced body weight, reduced energy expenditure, improved fasting insulin levels, and improved fatty acid and inflammatory biomarkers (low-density lipoprotein [LDL], total cholesterol-to-high-density lipoprotein [HDL] ratio, and C-reactive protein [CRP]). Reduced bone density, a potential problem, was not observed. A longer 2-year Phase II trial is in progress. Regarding life span, the authors of the CALERIE study estimated that on the basis of comparison with rodent data, a 55-year-old male beginning a DR regimen would gain an average of 2 months of additional life span(13–15).

Members of the Caloric Restriction Society (CRS) voluntarily limit total energy intake in the hopes of retarding aging. CRS members (predominately males with an average age 50 – 10 years who have undergone DR for an average of 6 years) when compared to age-matched controls consuming typical American diets, had a lower body mass index (BMI), reduced body fat, less total serum cholesterol, LDL cholesterol, total cholesterol/LDL, and higher HDL cholesterol. Fasting plasma insulin and glucose levels also were decreased for CRS members (16). Lower levels of plasma CRP and tumor necrosis factor-a (TNF-a) reflected decreased inflammation in CRS members. Circadian heart rate variability (HRV), a measure of autonomic function, was higher in CRS members, equivalent to people 20 years younger, while baseline average heart rate was lower. These data suggest rebalancing of the sympathetic-parasympathetic axis toward the parasympathetic in CRS members

(17). Left ventricular diastolic function in CRS members was similar to that found in people about 16 years younger (18). Taken together, CRS members appear to have substantially less risk for cardiovascular disease and diabetes. Although these data derive from a small cohort, they suggest that DR could have substantial health span benefit in humans.

Composition of diet

Does composition of diet matter in DR? Although the early dogma was that only total calories were important in DR, subsequent data suggest that this may not be true and that the composition of the DR diet does indeed matter. The variation in results in rhesus monkey DR studies may be partially due to differences in the amount of sucrose in their diet (see above). The type of DR may impact the effect of physical activity on DR (see above). Furthermore, in humans, unlike rodents, severe CR does not alter IGF-1 and IGF-1: IGF1 binding protein-3 (IGFBP-3) levels. In contrast, moderately reducing protein intake from 1.67 grams/kg to 0.95gram/kg reduces IGF-1 levels 27% (8). In humans, these data suggest that caloric-based DR may preferentially involve the mTOR/S6K pathway (**Figure 7.1**). Interestingly, DR via protein restriction extends life span in rodents, although not quite as much as restricting total calories, and in particular restriction of methionine appears to be sufficient to induce much of the benefit of protein restriction, perhaps by reducing oxidative damage to proteins and lowering mitochondrial reactive oxygen species (ROS) (19,20). It has been hypothesized that about 50% of the life extension effects of DR in rodents may be due to methionine restriction (19).

Health span benefits versus potential problems

Benefits

On the basis of clinical trials, potential health span benefits from DR in humans minimally include protection from diabetes and increased cardiovascular health. Moreover, data from the rodent and rhesus monkey studies suggest that DR may decrease the incidence of cancer as well, if these results translate to humans.

Potential drawbacks

DR reduces BMI and muscle mass (21). Although high BMI is considered unhealthy, low BMI is associated with increased mortality from all causes in middle-aged and elderly humans (22). Reduced muscle mass has been observed in DR humans. Loss of strength associated with reduced muscle mass may lower quality of life and conflict with the benefits of increased muscle mitochondrial biogenesis. Wound healing is diminished in DR-treated rats (23), which suggests that wound healing be carefully investigated in humans undergoing DR. Genetic background may be of great importance in humans as well as in the model systems. For example, there is some evidence that DR accelerates the course of neurodegeneration in a mouse model of amyotrophic lateral sclerosis (ALS) (24) which may contraindicate DR for people with a genetic predilection for ALS. DR may have paradoxical effects on immune function. On the one hand, DR has been reported to increase the number of naive T cells and the diversity of the T cell repertoire (which decrease in aging) as well as slowing the decline in antibody production, T cell proliferation, and antigen presentation in a variety of mammals (for review, see Spindler, 4). On the other hand, DR-treated mice have higher mortality after influenza infection, perhaps due to reduced reserves (20,25). Also, DR in old rhesus monkeys decreases T cell proliferation and can produce lymphopenia (26).

Alternatives

Should DR prove safe and effective in humans, adherence to a low-calorie diet may prove difficult for many. Therefore, it is useful to find alternate ways to achieve similar health span benefits. There are two approaches: The discovery of drugs that mimic DR and the possibility of modified DR which includes alternate day fasting (ADF), protein or methionine restriction, and combining a modest reduced-calorie diet with exercise.

Of the many reported therapeutics that increase longevity in rodents, rapamycin represents a rational choice (**Figure 7.1**) that inhibits key regulator mTOR and extended the life span of both male (10%) and female adult mice (13%) (7). However, unless and until a safe rapamycin analog is approved for human use that does not decrease glucose tolerance through inhibition of complex mTORC227 and lacks immunosuppressive activity, it can not be recommended for enhancing health span. Unfortunately, rapamycin's immunosuppressive capability is linked to its ability to bind mTOR, so a simple analog that elicits health span enhancement and is not immunosuppressive may be difficult.

The adenosine monophosphate-activated protein kinase (AMPK) agonist metformin (**Figure 7.1**) is an agent that appears capable of inducing some of the effects of DR; specifically it may prevent the onset of diabetes and reduce the incidence of cancer. Metformin is used in the treatment of type 2 diabetes; it increases glucose uptake in muscle and peripheral insulin sensitivity and decreases gluconeogenesis in the liver. It reproduces up to 75% of the genome-wide gene expression changes seen with DR in old mice (28) and extends life span in some mouse strains (between 8 and 38%) (29,30) but not rats (31). However, it has been hypothesized that metformin induces a mild DR effect by suppressing appetite that may explain the increased murine life span (4)

Alternate day fasting (ADF), in which subjects alternate fasting, with *ad libitum* feeding represents a more attractive way to achieve DR, because it imposes less stringent requirements on an individual. In rodents, ADF has been reported to achieve comparable results to conventional DR both in terms of life span extension and increased health span without body weight loss, although more well-controlled studies are needed (for review, see Varady, 32). Unfortunately, human ADF data is inadequate and contradictory reports about potential benefits do not allow for useful conclusions (33).

Even more intriguing is the possibility of using methionine restriction (MR) to achieve some of the effects of DR in humans, similar to those seen in rodents. Diets focused on fruits and vegetables can significantly lower methionine (34). However, human studies are necessary to confirm the validity of MR.

Combining a caloric-restricted diet with exercise (CE) would appear to be a particularly attractive way to achieve the health span benefits of DR, if the energy lost in exercise is balanced by restricting fewer calories. In animals and humans, CE generally achieves similar results to DR,(14,35–38) although in animals there appear to be few additional benefits beyond decreased loss of cardiac and skeletal muscle function in old age,(39,40) whereas in humans increased reduction in diastolic blood pressure, LDL cholesterol, and insulin sensitivity have been observed (38,41). In humans, epidemiological studies indicate that continued exercise correlates with an average of 2 years of additional life span in the elderly (42) At the same time, inactivity plays a major role in the secondary aging of essential physiological functions, suggesting exercise alone may be of benefit to human longevity (43).

Biomarkers

Finally, it would be of great value in evaluating the potential benefits of DR in humans to have better biomarkers for aging at the molecular and cellular levels to assess how much fundamental alteration of life span is really possible without SENS. Some biomarkers to consider in future DR studies include evaluation of the somatic cell mutation rate genome wide, changes in telomere length, epigenomic changes, and perhaps the development of new biomarkers, for example, protein aggregation.

Conclusions

To understand the maximum potential benefits from DR requires differentiating fundamental effects on aging at the cellular and molecular levels from suppression of age-associated diseases, such as cancer. To that end, it is important that the investigators carefully target the effects of DR on molecular aging biomarkers, such as mutation rate and epigenomic alterations. Furthermore, large, well-controlled human studies are needed to confirm the benefits of DR. A combination of modest balanced diet and exercise may be a good compromise for improved health span while waiting for specific therapeutics to slow aging and/or to promote regeneration/rejuvenation.

References

1. Mattison JA, Roth GS, Beasley TM, Tilmont EM, Handy AM, Herbert RL, Longo DL, Allison DB, Young JE, Bryant M, Barnard D, Ward WF, Qi W, Ingram DK, de Cabo R. Impact of caloric restriction on health and survival in rhesus monkeys from the NIA study. Nature 2012;13:318–321.
2. Colman RJ, Anderson RM, Johnson SC, Kastman EK, Kosmatka KJ, Beasley TM, Allison DB, Cruzen C, Simmons HA, Kemnitz JW, Weindruch R. Caloric restriction delays disease onset and mortality in rhesus monkeys. Science 2009; 325:201–204.
3. Katewa SD, Demontis F, Kolipinski M, Hubbard A, Gill MS, Perrimon N, Melov S, Kapahi P. Intramyocellular fatty-acid metabolism plays a critical role in mediating responses to dietary restriction in *Drosophila melanogaster*. Cell Metab 2012;16:97–103.
4. Spindler SR. Caloric restriction: From soup to nuts. Ageing Res Rev 2010;9:324–353.
5. McCay CM, Crowell MF, Maynard LA. The effect of retarded growth upon the length of life span and upon the ultimate body size. 1935. Nutrition 1989;5:155–171; discussion 172.
6. Fontana L, Partridge L, Longo VD. Extending healthy life span—from yeast to humans. Science 2010;328:321–326.
7. Harrison DE, Strong R, Sharp ZD, Nelson JF, Astle CM, Flurkey K, Nadon NL,

Wilkinson JE, Frenkel K, Carter CS, Pahor M, Javors MA, Fernandez E, Miller RA. Rapamycin fed late in life extends lifespan in genetically heterogeneous mice. Nature 2009;460:392–395.

8. Fontana L, Weiss EP, Villareal DT, Klein S, Holloszy JO. Long-term effects of calorie or protein restriction on serum IGF-1 and IGFBP-3 concentration in humans. Aging Cell 2008;7:681–687.

9. Swindell WR. Dietary restriction in rats and mice: A meta-analysis and review of the evidence for genotype-dependent effects on lifespan. Ageing Res Rev 2012;11:254–270.

10. Austad SN. Ageing: Mixed results for dieting monkeys. Nature 2012; doi:10(1)038/nature11484, published online 29 August 29, 2012.

11. Linford NJ, Chan TP, Pletcher SD. Re-patterning sleep architecture in drosophila through gustatory perception and nutritional quality. PLoS Genet 2012;8:e1002668.

12. Guevara-Aguirre J, Balasubramanian P, Guevara-Aguirre M, Wei M, Madia F, Cheng C-W, Hwang D, Martin-Montalvo A, Saavedra J, Ingles S, Cabo R de, Cohen P, Longo VD. Growth hormone receptor deficiency is associated with a major reduction in pro-aging signaling, cancer, and diabetes in humans. Sci Transl Med 2011;3:70ra13–70ra13.

13. Fontana L, Villareal DT, Weiss EP, Racette SB, Steger-May K, Klein S, Holloszy JO. Calorie restriction or exercise: Effects on coronary heart disease risk factors. A randomized, controlled trial. Am J Physiol Endocrinol Metab 2007;293: E197–E202.

14. Heilbronn LK, de Jonge L, Frisard MI, DeLany JP, Meyer DEL, Rood J, Nguyen T, Martin CK, Volaufova J, Most MM, Greenway FL, Smith SR, Williamson DA, Deutsch WA, Ravussin E. Effect of 6-mo. calorie restriction on biomarkers of longevity, metabolic adaptation and oxidative stress in overweight subjects. JAMA 2006;295:1539–1548.

15. Rochon J, Bales CW, Ravussin E, Redman LM, Holloszy JO, Racette SB, Roberts SB, Das SK, Romashkan S, Galan KM, Hadley EC, Kraus WE. Design and conduct of the calerie study: Comprehensive assessment of the long-term effects of reducing intake of energy. J Gerontol A Biol Sci Med Sci 2011;66A:97–108.

16. Fontana L, Meyer TE, Klein S, Holloszy JO. Long-term calorie restriction is highly effective in reducing the risk for atherosclerosis in humans. Proc Natl Acad Sci USA 2004; 101:6659–6663.

17. Stein PK, Soare A, Meyer TE, Cangemi R, Holloszy JO, Fontana L. Caloric restriction may reverse age-related autonomic decline in humans. Aging Cell 2012;11:644–650.

18. Holloszy JO, Fontana L. Caloric restriction in humans. Exp Gerontol 2007;42:709–712.

19. Pamplona R, Barja G. Mitochondrial oxidative stress, aging and caloric restriction: The protein and methionine connection. Biochim Biophys Acta

2006;1757:496–508.

20. Caro P, Gomez J, Sanchez I, Garcia R, Lopez-Torres M, Naudi A, Portero-Otin M, Pamplona R, Barja G. Effect of 40% restriction of dietary amino acids (except methionine) on mitochondrial oxidative stress and biogenesis, aif and sirt1 in rat liver. Biogerontology 2009;10:579–592.

21. Weiss EP, Racette SB, Villareal DT, Fontana L, Steger-May K, Schechtman KB, Klein S, Ehsani AA, Holloszy JO. Lower extremity muscle size and strength and aerobic capacity decrease with caloric restriction but not with exercise-induced weight loss. J Appl Physiol 2007;102:634–640.

22. Takata Y, Ansai T, Soh I, Akifusa S, Sonoki K, Fujisawa K, Awano S, Kagiyama S, Hamasaki T, Nakamichi I, Yoshida A, Takehara T. Association between body mass index and mortality in an 80-year-old population. J Am Geriatr Soc 2007;55:913–917.

23. Reiser K, McGee C, Rucker R, McDonald R. Effects of aging and caloric restriction on extracellular matrix biosynthesis in a model of injury repair in rats. J Gerontol A Biol Sci Med Sci 1995;50A:B40–B47.

24. Patel BP, Safdar A, Raha S, Tarnopolsky MA, Hamadeh MJ. Caloric restriction shortens lifespan through an increase in lipid peroxidation, inflammation and apoptosis in the g93a mouse, an animal model of als. PLoS ONE 2010; 5:e9386.

25. Ritz BW, Gardner EM. Malnutrition and energy restriction differentially affect viral immunity. J Nutr 2006;136: 1141–1144.

26. Messaoudi I, Fischer M, Warner J, Park B, Mattison J, Ingram DK, Totonchy T, Mori M, Nikolich-Zugich J. Optimal window of caloric restriction onset limits its beneficial impact upon t cell senescence in primates. Aging Cell 2008;7: 908–919.

27. Lamming DW, Ye L, Katajisto P, Goncalves MD, Saitoh M, Stevens DM, Davis JG, Salmon AB, Richardson A, Ahima RS, Guertin DA, Sabatini DM, Baur JA. Rapamycin-induced insulin resistance is mediated by mtorc2 loss and uncoupled from longevity. Science 2012; 335:1638–1643.

28. Dhahbi JM, Mote PL, Fahy GM, Spindler SR. Identification of potential caloric restriction mimetics by microarray profiling. Physiol. Genomics 2005;23:343–350.

29. Anisimov VN, Berstein LM, Popovich IG, Zabezhinski MA, Egormin PA, Piskunova TS, Semenchenko AV, Tyndyk ML, Yurova MN, Kovalenko IG, Poroshina TE. If started early in life, metformin treatment increases life span and postpones tumors in female shr mice. Aging (Albany NY) 2011; 3:148–157.

30. Anisimov VN, Berstein LM, Egormin PA, Piskunova TS, Popovich IG, Zabezhinski MA, Tyndyk ML, Yurova MV, Kovalenko IG, Poroshina TE, others. Metformin slows down aging and extends life span of female shr mice. Cell Cycle 2008;7:2769–2773.

31. Smith DL, Elam CF, Mattison JA, Lane MA, Roth GS, Ingram DK, Allison DB. Metformin supplementation and life span in Fischer-344 rats. J Gerontol A Biol Sci Med Sci 2010; 65A:468–474.

32. Varady KA, Hellerstein MK. Alternate-day fasting and chronic disease

prevention: A review of human and animal trials. Am J Clin Nutr 2007;86:7–13.
33. Trepanowski JF, Canale RE, Marshall KE, Kabir MM, Bloomer RJ. Impact of caloric and dietary restriction regimens on markers of health and longevity in humans and animals: A summary of available findings. Nutr J 2011;10:107.
34. McCarty MF, Barroso-Aranda J, Contreras F. The low methionine content of vegan diets may make methionine restriction feasible as a life extension strategy. Med Hypotheses 2009;72:125–128.
35. Holloszy JO. Mortality rate and longevity of food-restricted exercising male rats: A reevaluation. J Appl Physiol 1997;82: 399–403.
36. Civitarese AE, Carling S, Heilbronn LK, Hulver MH, Ukropcova B, Deutsch WA, Smith SR, Ravussin E. Calorie restriction increases muscle mitochondrial biogenesis in healthy humans. PLoS Med 2007;4.
37. Larson-Meyer DE, Newcomer BR, Heilbronn LK, Volaufova J, Smith SR, Alfonso AJ, Lefevre M, Rood JC, Williamson DA, Ravussin E. Effect of 6-month calorie restriction and exercise on serum and liver lipids and markers of liver function. Obesity (Silver Spring) 2008;16:1355–1362.
38. Lefevre M, Redman LM, Heilbronn LK, Smith JV, Martin CK, Rood JC, Greenway FL, Williamson DA, Smith SR, Ravussin E. Caloric restriction alone and with exercise improves CVD risk in healthy non-obese individuals. Atherosclerosis 2009;203:206–213.
39. Abete P, Testa G, Galizia G, Mazzella F, Della Morte D, de Santis D, Calabrese C, Cacciatore F, Gargiulo G, Ferrara N, Rengo G, Sica V, Napoli C, Rengo F. Tandem action of exercise training and food restriction completely preserves ischemic preconditioning in the aging heart. Exp Gerontol 2005;40:43–50.
40. Horska A, Brant LJ, Ingram DK, Hansford RG, Roth GS, Spencer RGS. Effect of long-term caloric restriction and exercise on muscle bioenergetics and force development in rats. Am J Physiol Endocrinol Metab 1999;276:E766–E773.
41. Larson-Meyer DE, Redman L, Heilbronn LK, Martin CK, Ravussin E. Caloric restriction with or without exercise: The fitness vs. fatness debate. Med Sci Sports Exerc 2010;42:152–159.
42. Rizzuto D, Orsini N, Qiu C, Wang H-X, Fratiglioni L. Life-style, social factors, and survival after age 75: Population based study. Br Med J 2012;345:e5568–e5568.
43. Booth FW, Laye MJ, Roberts MD. Lifetime sedentary living accelerates some aspects of secondary aging. J Appl Physiol 2011;111:1497–1504.

Chapter 8

Fibroblast growth factor-21 is a promising dietary restriction mimetic

Dietary or caloric restriction (DR or CR), typically a 30%–40% reduction in ad libitum or "normal" nutritional energy levels, has been reported to extend life span and health span in diverse organisms, including mammals. Although the life span benefit of DR in primates and humans is unproven, preliminary evidence suggests that DR confers health span benefits. A serious effort is underway to discover or engineer DR mimetics. The most straightforward path to a DR mimetic requires a detailed understanding of the molecular mechanisms that underlie DR and related life span–enhancing protocols. Increased expression of fibroblast growth factor-21 (FGF21), a putative mammalian starvation master regulator, promotes many of the same beneficial physiological changes seen in DR animals, including decreased glucose levels, increased insulin sensitivity, and improved fatty acid/lipid profiles. Ectopic overexpression of FGF21 in transgenic mice (FGF21-Tg) extends life span to a similar extent as DR in a recent study. FGF21 may achieve these effects by attenuating growth hormone (GH)/insulin-like growth factor-1 (IGF1) signaling. Although FGF21 expression does not increase during DR, and therefore is unlikely to mediate DR, it does increase during short-term starvation in rodents, which is a critical component of alternate day fasting, a DR-like protocol that also increases life span and health span in mammals. Various drugs have been reported to induce FGF21, including peroxisome proliferator-activated receptor-a (PPARa) agonists such as fenofibrate, the histone deacetylase inhibitor sodium butyrate, and adenosine monophosphate (AMP) kinase activators metformin and 5-amino-1-b-D-ribofuranosyl-imidazole-4-carboxamide (AICAR). Of these, only metformin has been reported to extend life span in mammals, and the extent of benefit is less than that seen with ectopic FGF21 expression. Perhaps the most parsimonious explanation is that high, possibly un-physiological, levels of FGF21 are needed to achieve maximum life span and health span benefits and that sufficiently high levels are not achieved by the identified FGF21 inducers. More in-depth studies of the effects of FGF21 and its inducers on longevity and health span are warranted.

Introduction

Dietary or caloric restriction (DR/CR), typically a 30%–40% reduction in ad libitum or "normal" nutritional + energy levels, has been reported to extend life span and health span in diverse organisms. Although mammalian data have been somewhat inconsistent, and convincing human benefit remains unestablished (1), a serious effort is underway to discover or engineer DR mimetics to avoid the obvious problems with practical implementation of DR or CR. The most straightforward path toward a DR mimetic requires a detailed understanding of the molecular mechanisms that underlie DR. DR effects can be achieved a variety of ways including inhibiting mammalian target of rapamycin (mTOR) with rapamycin or suppressing the IIS (insulin/ insulin-like growth factor [IGF-1]) pathways by mutagenesis. However, beyond the possible use of rapamycin as a DR mimetic (for review, see refs. 1 and 2), few potential DR mimetics have been identified.

It has been hypothesized that agents that suppress the IIS pathways, especially GH/IGF-1 in mammals would act as DR mimetics. One such agent may be fibroblast growth factor-21 (FGF21), a master regulator of the homeostatic response to starvation that inhibits IGF-1/growth hormone (GH) signaling (3) FGF21 is an atypical fibroblast growth factor that does not bind heparin. It functions as an endocrine hormone and also acts via autocrine mechanisms. FGF21 signals by binding a complex of beta-klotho/fibroblast growth factor receptor 1 (FGFR1) at the cell surface

(**Figure 8.1**). In liver, FGF21 inhibits the GH/IGF-1 signaling pathway by blocking Janus kinase 2 (JAK2)-mediated phosphorylation and nuclear translocation of the transcription factor, signal transducer and activator of transcription 5 (STAT5) (4–6). This, in turn, suppresses the transcription of IGF-1 and other GH/ STAT5-regulated genes (5). FGF21 also induces the IGF-1 binding protein-1 and the suppressor of cytokine signaling 2, which further attenuates GH/IGF-1 signaling (6). Thus, FGF21-mediated repression of the GH/IGF-1 axis provides a mechanism for blocking growth and conserving energy under starvation conditions (4). These activities suggest that FGF21 may be able to confer many of the benefits of DR.

Ectopic Expression of FGF21 Extends Life Span

In a potentially important paper, Zhang *et al.* show that ectopic expression of FGF21 in transgenic mice extends mean life span 36% without reducing food intake (70. Despite having similar levels of food intake to wild-type mice, FGF21-overexpressing mice achieved many of the benefits associated with DR, including high insulin sensitivity. The transgenic mice (FGF21-Tg) express FGF21 selectively in hepatocytes using the strong apolipoprotein E (apoE) promoter, which results in serum FGF21 levels 5- to 10-fold higher than that achieved by starvation in wild-type mice (7). Decreases in serum insulin, IGF-1, glucose, triglycerides, and cholesterol have been observed in young FGF21-Tg mice (5, 8) that are consistent with the known effects of FGF21 signaling. Old (26- to 27-month) FGF21-Tg mice have lower plasma IGF-1, fasting plasma glucose levels, and lower insulin levels in response to challenge with glucose. FGF21-Tg mice have higher insulin sensitivity, as demonstrated by hyperinsulinemic–euglycemic clamp studies that show that a higher glucose infusion rate is required to maintain euglycemia in TGF21-Tg mice compared with wild-type mice. Furthermore, glucose tracer kinetics showed that in FGF21-Tg mice insulin suppressed hepatic glucose production and increased whole-body disposal of glucose to a greater degree than in wild-type mice.

Physically, FGF21-Tg mice are not only similar to wildtype mice in the amount of food they eat, but also in their levels of physical activity, respiration, oxygen consumption, and percent fat and lean body mass. The most overt physical difference observed is that FGF21-Tg mice are smaller than wild-type mice, which the authors hypothesize is due to impaired GH/IGF-1 signaling resulting from lower levels of the downstream GH mediator IGF-1. IGF-1 levels were lower in spite of elevated GH plasma levels Consistent with the unchanged body mass parameters, adiponectin and leptin levels are unchanged in females. Although maximum life span has not yet been determined, FGF21-Tg females lived longer than males—greater than 30% of mice remain alive at 44 months. Interestingly, females also had significantly lower IGF-1 levels than males. No alterations in pathways associated with other known longevity regulators such as mTOR, adenosine monophosphate (AMP) kinase, and sirtuin/nicotine adenine dinucleotide (NAD) were observed in preliminary experiments (7).

Because FGF21 over-expression seemed to mimic some of the critical effects seen in long-lived dwarf mice carrying mutations in GH signaling and mice subjected to DR, the authors performed microarray analysis to investigate RNA expression of 43 previously identified candidate longevity associated genes. Liver expression of longevity candidate genes Fmo3, Igfals, Hes6, Alas2, Cyp4a12b, Mup4, Serpina12, and Hsd3b5 are similarly co-regulated by FGF21 over- expression and caloric restriction. Because FGF21 signals through a STAT5-dependent mechanism, the authors hypothesize that the physiological and longevity effects of FGF21 may also be STAT5 dependent (**Figure 8.1**). Consistent with this hypothesis, 26 of the 33 candidate longevity genes altered by FGF21 over-expression were previously observed to be regulated similarly in STAT5-deficient mice,9 leading Zhang *et al.* to hypothesize that DR and FGF21 over-expression may both increase longevity by

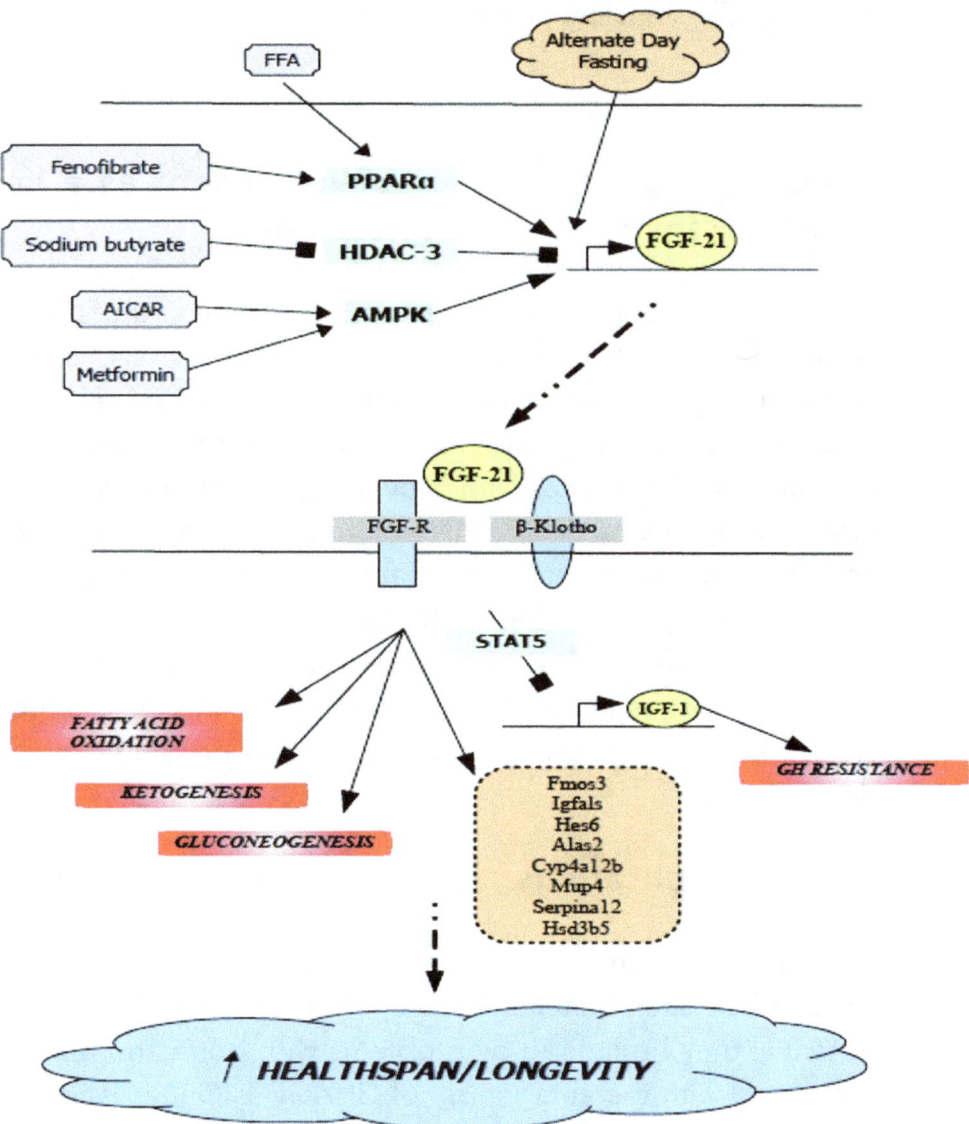

Figure 8.1 Fibroblast growth factor (FGF) signaling and health span. (Top) Hepatocytes induce FGF21 expression in response to starvation (alternate day fasting in rodents), fenofibrate (via activation of peroxisome proliferator-activated receptor-α [PPARα]), sodium butyrate (via inhibition of histone deacetylase 3 [HDAC-3]), 5-amino-1-β-D-ribofuranosyl-imidazole-4-carboxamide (AICAR), and metformin (via activation of adenosine monophosphate kinase [AMPK]). (Bottom) FGF21 binds to target cells at beta-klotho/FGF receptors causing stimulation of fatty acid oxidation, ketogenesis, gluconeogenesis and inhibition of growth hormone (GH)/insulin-like growth factor-1 (IGF-1) signaling through reduction of signal transducer and activator of transcription-5 (STAT5). FGF21 signaling leads to increased health span/longevity. FFA, Free fatty acids.

inhibiting GH signaling pathways in liver (7).

Is FGF21 a candidate for development as a health span/ life span–enhancing hormone?

Clearly, replication of these studies, including assessing the ability of FGF21 treatment to benefit adult animals, is needed. Furthermore, the authors observe that FGF21-Tg mice have lower bone density than wild-type mice. Bone loss and increased risk of fracture have been observed previously in studies evaluating the effects of exogenous FGF21 as a potential treatment for diabetes and are associated with potentiation of peroxisome proliferator activated receptor c (PPARc)(10). These results are not surprising given that PPARc is known to oppose insulin/IGF-1 signaling11 and IGF-1 signaling is important to normal bone and muscle function. However, these results present a serious problem for clinical development.

Medical Implications

FGF21 is an intriguing DR mimetic that may confer health span and life span benefits that merit further study. However, clinical potential may be limited by problems with bone and muscle loss associated with the antagonism of GH/IGF-1. One possible way to overcome this problem would be to find an agent that induces FGF21 and also possesses activities that reduce or eliminate potential bone and muscle deterioration.

The simplest way to induce FGF21 is by short-term starvation. Interestingly, alternate day fasting (ADF), a form of DR that involves alternating a day of fasting with ad libitum feeding has already been reported to confer many of the benefits of DR (for review, see ref. 12). We hypothesize that FGF21 is likely to play an important role in achieving the physiological benefits of ADF in rodents (**Figure 8.1**) and that ADF, at least in rodents, may be somewhat different from DR at the mechanistic level, given that FGF21 levels do not increase during chronic CR or DR in the mammals investigated so far. Although 24 hr is sufficient to induce FGF21 in mice, 7 days are necessary to increase plasma FGF21 levels in humans,13 which suggests that 24-hr ADF protocols may not be as efficient in humans as in rodents if FGF21 plays a critical role in both. The issue of bone and muscle loss may be dose dependent, and FGF21 levels during starvation are significantly less (5- to 10-fold lower) than in the FGF21-Tg transgenic mice, which may explain the apparent minimal damage to bone homeostasis in ADF. Careful studies evaluating the role of FGF21 levels in ADF are warranted with special emphasis on the role of IGF-1 function and bone/ muscle loss.

PPAR alpha agonists such as fenofibrate are also known to strongly induce liver-specific expression of FGF21 (**Figure 8.1**)(13-14). FGF21 may be the key downstream target of PPARa in mediating responses to starvation, including causing changes in ketogenesis, fatty acid oxidation, and gluconeogenesis. Interestingly, fenofibrate is an approved drug for hypercholesterolemia and hypertriglyceridemia. Fenofibrate's primary mechanism of action is to activate lipoprotein lipase and reduce apolipoprotein CIII, increasing high-density lipoprotein (HDL) and decreasing low-density lipoprotein (LDL) and very-low-density lipoprotein (VLDL). That effects on sugar metabolism are not observed signifies that FGF21 is not the sole downstream target of PPARa or that FGF21 plasma levels do not reach a threshold needed for suppression of GH/IGF-1. A potential role of fenofibrate in prolongation of longevity is unknown, but should be investigated.

Sodium butyrate, an inhibitor of histone deacetylase 3, increases plasma levels and liver expression of FGF21 in mice15 (**Figure 8.1**). Interestingly, sodium butyrate has protective effects against obesity and dyslipidemia as well as helping to restore age-related memory loss in rats (16) In *Drosophila melanogaster*, sodium butyrate increased life span when administered to old animals, but it decreased life span when administered over the entire health span (17) Whether sodium butyrate has any value as an inducer of FGF21 and for health span enhancement requires further study.

The AMP kinase activators metformin and 5-amino-1-b-D-ribofuranosyl-imidazole-4-carboxamide (AICAR) also have been reported to increase FGF21 expression in liver cells(1)8 That metformin raises FGF21 levels is intriguing because it has been reported to possess many of the properties attributed to over-expression of FGF21 (**Figure 8.1**). Metformin is used as first-line treatment of type 2 diabetes because it increases glucose uptake in muscle and peripheral insulin sensitivity and decreases gluconeogenesis in the liver. Metformin reproduces up to 75% of the genome-wide gene expression changes seen with DR in old mice (19), and extends life span in some mouse strains (between 8 and 38%) (20-21). Furthermore, metformin may prevent the onset of diabetes and reduce the incidence of cancer. It would be very tempting to hypothesize that many of the beneficial effects of metformin are mediated through induction of FGF21, and this idea should be carefully tested. Metformin does not cause bone loss and that may be due to activated AMP kinase promoting bone formation (22-23). Metformin is a Food and Drug Administration (FDA)-approved drug for diabetes and may be the safest of the potential DR or ADF mimetics available.

AICAR also activates AMP kinase, but it is less well studied than metformin and appears to lack many of the positive effects of metformin on blood sugar. Given its ability to induce FGF21, AICAR should be investigated for similar life span/ health span benefits. AICAR is a particularly intriguing potential DR mimetic that has been reported to increase osteoblast differentiation (22) and to act as an exercise mimetic (24-25). Whether any of these inducers are potent enough inducers of FGF21 to achieve health span benefit is an open question. FGF21-Tg animals exhibit plasma levels of FGF21 that are up to 10-fold greater than the levels observed with starvation. Such levels may well be un-physiological, and not realizable by conventional induction, especially given the instability of plasma FGF21 (26-27).

Conclusions

DR or CR has been reported to extend life span and health span in diverse organisms, including mammals. Although the life span benefit of DR in primates and humans is unproven, preliminary evidence suggests that DR confers health span benefits, especially relating to blood sugar, lipids, and cardiovascular health. The quest for DR mimetics is motivated by the fact that typical DR or even ADF regimens are impractical for most people. Over-expression of FGF21, a putative mammalian starvation master regulator, extends life span in mice and improves insulin sensitivity and fatty acid/lipid profiles, raising the possibility that this cytokine may be a good DR mimetic. Unfortunately, even should these pre-clinical results be replicated, translation to the clinic is unlikely given the potential for bone loss. However, compounds that induce sufficient FGF21 for health span benefit without bone or muscle loss may be found. Metformin may be one such compound.

References

1. Mendelsohn AR, Larrick JW. Dietary restriction: Critical co-factors to separate health span from life span benefits. Rejuvenation Res 2012;15:523–529.
2. Mouchiroud L, Molin L, Dalliere N, Solari F. Life span extension by resveratrol, rapamycin, and metformin: The promise of dietary restriction mimetics for healthy aging. Biofactors 2010;36:377–382.
3. Kliewer SA, Mangelsdorf DJ. Fibroblast growth factor 21: From pharmacology to physiology. Am J Clin Nutr 2010;91: 254S–257S.
4. Ge X, Wang Y, Lam KS, Xu A. Metabolic actions of FGF21: Molecular mechanisms and therapeutic implications. Acta Pharmaceutica Sinica B

2012;2:350–357.

5. Inagaki T, Lin VY, Goetz R, Mohammadi M, Mangelsdorf DJ, Kliewer SA. Inhibition of growth hormone signaling by the fasting-induced hormone FGF21. Cell Metab 2008;8: 77–83.

6. Kubicky RA, Wu S, Kharitonenkov A, De Luca F. Role of fibroblast growth factor 21 (FGF21) in undernutrition-related attenuation of growth in mice. Endocrinology 2012; 153:2287–2295.

7. Zhang Y, Xie Y, Berglund ED, Coate KC, He TT, Katafuchi T, Xiao G, Potthoff MJ, Wei W, Wan Y. The starvation hormone, fibroblast growth factor-21, extends lifespan in mice. eLife 2012;1.

8. Inagaki T, Dutchak P, Zhao G, Ding X, Gautron L, Parameswara V, Li Y, Goetz R, Mohammadi M, Esser V, Elmquist JK, Gerard RD, Burgess SC, Hammer RE, Mangelsdorf DJ, Kliewer SA. Endocrine regulation of the fasting response by PPARalpha-mediated induction of fibroblast growth factor 21. Cell Metab 2007;5:415–425.

9. Barclay JL, Nelson CN, Ishikawa M, Murray LA, Kerr LM, McPhee TR, Powell EE, Waters MJ. GH-dependent STAT5 signaling plays an important role in hepatic lipid metabolism. Endocrinology 2011;152:181–192.

10. Wei W, Dutchak PA, Wang X, Ding X, Wang X, Bookout AL, Goetz R, Mohammadi M, Gerard RD, Dechow PC, Mangelsdorf DJ, Kliewer SA, Wan Y. Fibroblast growth factor 21 promotes bone loss by potentiating the effects of peroxisome proliferator-activated receptor c. Proc Natl Acad Sci USA 2012;109:3143–3148.

11. Belfiore A, Genua M, Malaguarnera R. PPAR-c agonists and their effects on IGF-I receptor signaling: Implications for cancer. PPAR Res 2009;2009.

12. Trepanowski JF, Canale RE, Marshall KE, Kabir MM, Bloomer RJ. Impact of caloric and dietary restriction regimens on markers of health and longevity in humans and animals: A summary of available findings. Nutr J 2011;10:107.

13. Badman MK, Pissios P, Kennedy AR, Koukos G, Flier JS, Maratos-Flier E. Hepatic fibroblast growth factor 21 is regulated by PPARalpha and is a key mediator of hepatic lipid metabolism in ketotic states. Cell Metab 2007;5:426–437.

14. Galman C, Lundasen T, Kharitonenkov A, Bina HA, Eriksson M, Hafstro m I, Dahlin M, Amark P, Angelin B, Rudling M. The circulating metabolic regulator FGF21 is induced by prolonged fasting and PPARalpha activation in man. Cell Metab 2008;8:169–174.

15. Li H, Gao Z, Zhang J, Ye X, Xu A, Ye J, Jia W. Sodium butyrate stimulates expression of fibroblast growth factor 21 in liver by inhibition of histone deacetylase 3. Diabetes 2012;61:797–806.

16. Reolon GK, Maurmann N, Werenicz A, Garcia VA, Schro der N, Wood MA, Roesler R. Posttraining systemic administration of the histone deacetylase inhibitor sodium butyrate ameliorates aging-related memory decline in rats. Behav Brain

Res 2011;221:329–332.
17. McDonald P, Maizi BM, Arking R. Chemical regulation of mid- and late-life longevities in drosophila. Exp Gerontol 2012; http://dx.doi.org/10(1)016/j.exger.2012.09.006/.
18. Nygaard EB, Vienberg SG, Ørskov C, Hansen HS, Andersen B. Metformin stimulates FGF21 expression in primary hepatocytes. Exp Diabetes Res 2012;2012:1–8.
19. Dhahbi JM, Mote PL, Fahy GM, Spindler SR. Identification of potential caloric restriction mimetics by microarray profiling. Physiol Genomics 2005;23:343–350.
20. Anisimov VN, Berstein LM, Popovich IG, Zabezhinski MA, Egormin PA, Piskunova TS, Semenchenko AV, Tyndyk ML, Yurova MN, Kovalenko IG, Poroshina TE. If started early in life, metformin treatment increases life span and postpones tumors in female SHR mice. Aging (Albany NY) 2011;3:148–157.
21. Anisimov VN, Berstein LM, Egormin PA, Piskunova TS, Popovich IG, Zabezhinski MA, Tyndyk ML, Yurova MV, Kovalenko IG, Poroshina TE, others. Metformin slows down aging and extends life span of female SHR mice. Cell Cycle 2008;7:2769–2773.
22. Jang WG, Kim EJ, Bae I-H, Lee K-N, Kim YD, Kim D-K, Kim S-H, Lee C-H, Franceschi RT, Choi H-S, Koh J-T. Metformin induces osteoblast differentiation via orphan nuclear receptor SHP-mediated transactivation of runx2. Bone 2011;48: 885–893.
23. Jeyabalan J, Shah M, Viollet B, Chenu C. AMP-activated protein kinase pathway and bone metabolism. J Endocrinol 2011;212:277–290.
24. Narkar VA, Downes M, Yu RT, Embler E, Wang Y-X, Banayo E, Mihaylova MM, Nelson MC, Zou Y, Juguilon H, Kang H, Shaw RJ, Evans RM. AMPK and PPARdelta agonists are exercise mimetics. Cell 2008;134:405–415.
25. Bueno Junior CR, Pantaleao LC, Voltarelli VA, Bozi LHM, Brum PC, Zatz M. Combined effect of AMPK/PPAR agonists and exercise training in mdx mice functional performance. PLoS One 2012;7:e45699.
26. Xu J, Stanislaus S, Chinookoswong N, Lau YY, Hager T, Patel J, Ge H, Weiszmann J, Lu S-C, Graham M, Busby J, Hecht R, Li Y-S, Li Y, Lindberg R, Veniant MM. Acute glucose-lowering and insulin-sensitizing action of FGF21 in insulin-resistant mouse models—association with liver and adipose tissue effects. Am J Physiol Endocrinol Metab 2009;297:E1105–E1114.
27. Huang Z, Wang H, Lu M, Sun C, Wu X, Tan Y, Ye C, Zhu G, Wang X, Cai L, Li X. A better anti-diabetic recombinant human fibroblast growth factor 21 (rhFGF21) modified with polyethylene glycol. PLoS One 2011;6:e20669.

Chapter 9

Dietary modification of the microbiome affects risk for cardiovascular disease

The incidence of cardiovascular disease (CVD) increases with age and is associated with some syndromes that exhibit aspects of premature aging, such as progeria. Various factors are thought to contribute to the progression of CVD, including hypertension, hypercholesterolemia, diets rich in saturated and trans fats, etc. Recent reports have uncovered an important connection between diet, the microbiome, and CVD. Dietary carnitine (present predominately in red meat) and lecithin (phosphatidyl choline) are shown to be metabolized by gut microbes to trimethylamine (TMA), which in turn is metabolized by liver flavin monoxygenases (especially FMO3 and FMO1) to form trimethylamine-N-oxide (TMAO). High levels of TMAO in the blood strongly correlate with CVD and associated acute clinical events. Plasma TMAO levels may be an important clinical biomarker for CVD. The data suggest that that presence of specific as yet unidentified microorganisms in the gut linked to diet are required for high TMAO levels and TMAO-mediated CVD progression. Development of novel therapeutic approaches to manipulate gut flora may help treat CVD.

Introduction

The incidence of cardiovascular disease (CVD) increases with age and is associated with some syndromes that exhibit aspects of premature aging, such as progeria. The American Heart Association estimates that as many as 83% of men and 87% of women greater than 80 years of age suffer from some form of CVD. Atherosclerosis is a multifaceted, progressive, inflammatory disease affecting mainly large and medium-sized arteries by build-up of so-called "atherosclerotic plaques." These lesions are comprised of a well-defined structure of lipids, necrotic cores, calcified regions, inflamed smooth muscle cells, endothelial cells, immune cells, and foam cells. Macrophages, derived from monocytes, play a pivotal role in the development and progression of atherosclerosis (1). A key event of atherosclerosis involves the uncontrolled uptake of oxidized low-density lipoproteins (oxLDL) in macrophages, which accumulate within the sub-endothelial space of blood vessel walls.2 Macrophages characterized by high expression of both mannose and CD163 receptors preferentially traffic to atherosclerotic lesions at sites of intra-plaque hemorrhage (3). When macrophages fail to restore their cellular cholesterol homeostasis via reverse cholesterol transport (RCT), they transform into foam cells, the main components of fatty streaks. In a complex series of steps still being elucidated, fatty streaks develop into atheromas.

Inflammatory processes mediated by cell–cell interactions in atheromas eventually lead to the death of the initial foam cells and macrophages, the recruitment and differentiation of new macrophages and foam cells, and the development of mature atherosclerotic plaques. Atherosclerotic plaques eventually rupture, providing a nidus for platelet activation, thrombus formation, and vessel obliteration resulting in myocardial infarction (MI) if the lesions enter coronary circulation. Various factors are thought to contribute to the progression of CVD, including hypertension, hypercholesterolemia, diets rich in saturated and trans fats, etc. However, the discovery of new co-factors may have great benefit for development of a cure or a preventive regiment for CVD.

Serum Trimethylamine-N-Oxide Levels Are Linked to CVD and Result from Microbiome-Mediated Metabolism of Dietary Carnitine and Lecithin

In an important series of recent papers from Stanley Hazen's laboratory (Cleveland Clinic), dietary carnitine and lecithin (phosphatidyl choline) are shown to be metabolized by gut microbes to trimethylamine (TMA), which in turn is metabolized by liver flavin monoxygenases (especially FMO3 and FMO1) (4) to form trimethylamine-N-oxide (TMAO). High levels of TMAO in the blood strongly correlate with CVD and acute clinical events (5–7) These are among the first reports to demonstrate a potential powerful connection between diet, the microbiome, and human disease.

The first clue that TMAO levels may play a critical role in CVD came with metabolic magnetic resonance imaging (MRI) and mass spectrometry studies on plasma from 40 samples selected from 2000+ cardiac patients. Those who had experienced a MI, stroke, or death during a 3-year period were compared with matched controls. TMAO was identified as the major analyte correlating with CVD (7). TMAO is an oxidation product of TMA, which was known to be formed by microorganisms from phosphatidylcholine (PC), choline, and other TMA-containing species, such as betaine (8). Fasting levels of 1,876 stable subjects showed a correlation of elevated TMAO, choline, and betaine with CVD and CVD phenotypes, such as peripheral artery disease, coronary artery disease, and MI (7).

Wang *et al.* (7) showed that gut flora have an obligate role in TMAO formation from dietary choline, using deuterated PC to trace the formation of TMAO in mice pretreated with broad-spectrum antibiotics for 3 weeks versus untreated control animals. TMAO was formed only in the negative control animals, but not the antibiotics-treated mice. As expected, exposing antibiotics-pre-treated mice to normal animals for 4 weeks subsequent to antibiotic treatment to allow microbe recolonization reconstituted the ability of the animals to form TMAO. The metabolic pathway from PC to TMAO was shown to be PC choline/TMA/TMAO (**Figure 9.1**) (7).

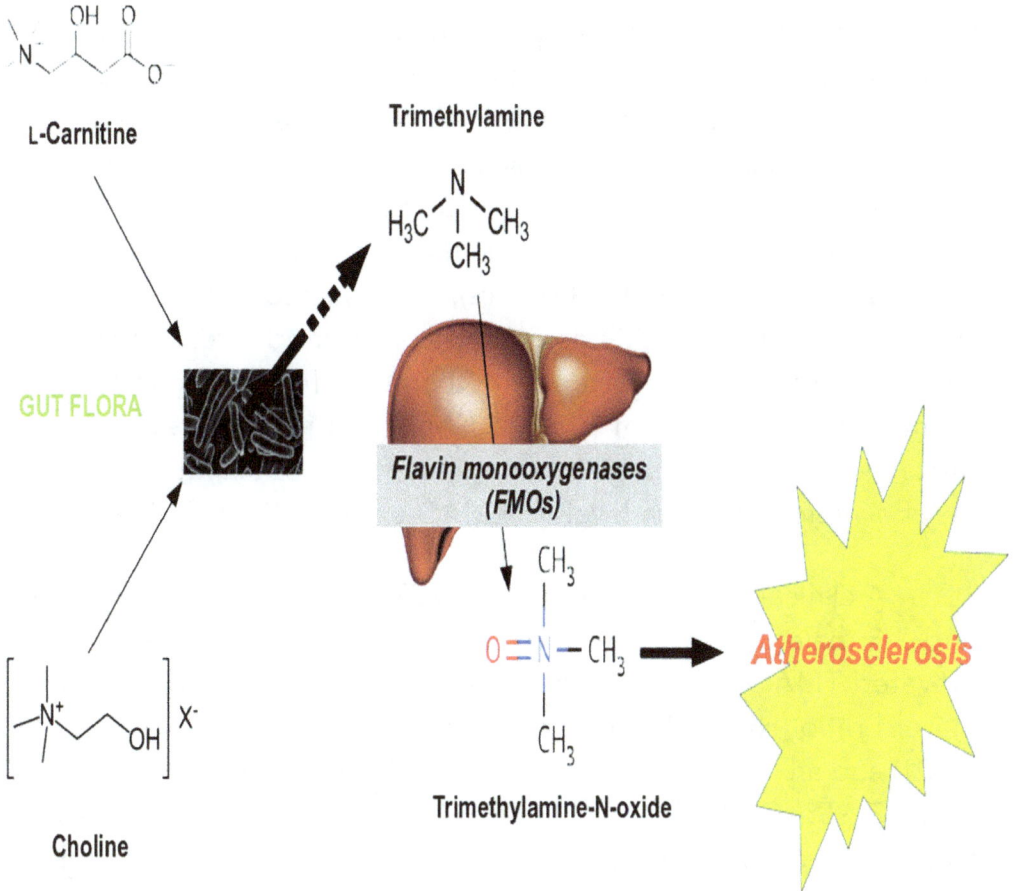

Figure 9.1 Trimethylamine-N-oxide (TMAO) metabolism and cardiovascular disease.

Dietary choline or TMAO supplementation enhanced atherosclerotic lesion development in atherosclerotic-prone C57BL/6J.Apoe-/- mice. Plasma TMAO levels were strongly correlated with plaque size, whereas blood lipid levels did not increase. Increased plaque formation could be abrogated by co-treatment with antibiotics, consistent with the need for microbial metabolism. Macrophage scavenger markers CD36 and SR-A1 as well as cholesterol-laden foam cell formation increased in C57BL/6J.Apoe-/- mice fed high-choline diets, effects that were dependent on the presence of normal gut flora (7). These results were not surprising given the increase in plaque formation and the important role macrophages and macrophage-derived foam cells play in atherosclerosis.

To confirm the clinical relevance of the mouse results, two clinical studies in humans were performed. Blood and urine samples were analyzed from 40 human volunteers treated with a mixture of unlabeled and deuterium-labeled PC. Six of the volunteers were given metronidazole and ciprofloxacin for a week and retested. The volunteers were then retested at 1 month. TMAO and TMA were detected after choline stimulation in the first visit and at 1 month (3 weeks after withdrawal of antibiotics). As expected from the mouse studies, almost no TMAO or TMA was detected after 1 week of antibiotic treatment, consistent with the obligate role of gut flora in humans in induction of plasma TMAO from choline.5 Some protection from possible pathological effects of elevated TMAO may be afforded by high levels of TMAO excretion in the urine.

A clinical outcomes study on 4,007 patients (average age 63 and 2/3 male) at the Cleveland Clinic revealed that subjects with elevated TMAO (highest quartile) had an increased risk of death (hazard ratio [HR], 3.7; 95% confidence interval [CI], 2.9–4.75; $p < 0.001$) and an increased nonfatal myocardial infarction or stroke rate (HR, 2.3; 95% CI, 1.48–3.05; $p<0.001$) (5). Overall inclusion of TMAO as a co-variate with known risk factors improved risk estimation by 8.6% (5). These data suggest that determining TMAO levels may prove to an effective biomedical marker for major event risk in CVD.

Given the potential importance of TMAO to the pathogenesis of CVD and the recent meta-analysis of prospective cohort studies showing no association of CVD with dietary saturated fat intake (9), Koeth et al. (6) examined if L-carnitine, a TMA abundant in red meat with a similar structure to choline (**Figure. 9.1**), also resulted in elevated TMAO levels and increased atherosclerosis in experiments that parallel those done for PC. A reanalysis of the unbiased small-metabolite metabolomic analysis that identified TMAO as a CVD risk revealed that L-carnitine was also associated with CVD. Human volunteers fed an 8-ounce steak and deuterated L-carnitine showed modestly increased TMAO and L-carnitine over time. Five volunteers were treated with broad-spectrum antibiotics for a week, which completely abrogated the increase in plasma TMAO, but not the increase in carnitine. Re-challenge with carnitine several weeks after the cessation of antibiotic treatment again resulted in elevated TMAO, suggesting that intestinal microbiota are necessary for carnitine to increase TMAO levels, similar to the results for choline challenge (6).

High variation in the response to dietary carnitine and the importance of gut flora suggested that the subjects' dietary habits should be analyzed. Long-term vegans (>5 years) were compared with omnivores who were frequent red meat eaters after a carnitine challenge. Both fasting TMAO levels and carnitine-induced TMAO levels were markedly reduced in the vegans compared with the meat-eating omnivores. Analysis of bacterial 16S RNA sequences in fecal specimens showed that on average individuals with enriched bacteria of the genus Prevotella had higher TMAO levels than those with an enrichment of the genus *Bacteroides* (6). However, more work needs to performed to identify the specific microorganisms involved in metabolizing carnitine and PC to TMA.

As expected, mice fed carnitine responded similarly to those challenged with choline (see above): Carnitine-induced elevated TMAO that was dependent upon the presence of intact gut flora and promoted atherosclerotic plaque formation in Apoe -/- mice. Interestingly, 16S analysis in mice showed different bacterial taxa associated with high TMAO levels than those seen in humans. This was not entirely surprising given known differences in murine and human flora (10). These results show both the applicability and limits of mouse models to human pathology involving the microbiome and should inform other studies dependent upon bacterial-mediated pathology in mice.

A more extensive series of studies to elucidate potential mechanisms by which TMAO promotes atherosclerosis showed that:

(1) TMAO does not compete with arginine to reduce nitric oxide synthesis;

(2) TMAO increases the rate of cholesterol influx via increased expression of scavenger receptors CD36 and SRA (see above);

(3) TMAO does not increase the rate of cholesterol synthesis in macrophages;

(4) TMAO does not increase macrophage inflammatory gene expression or desmosterol;

(5) TMAO does inhibit RCT, which measures the efficiency of cholesterol removal from macrophages, but expression of many known cholesterol transporters did not change in a way consistent with reduced RCT, leaving the precise mechanism unknown; and

(6) TMAO decreased the bile acid pool size and lowered the expression of key bile acid synthesis (Cyp7a1, Cyp27a1) and transport proteins in the liver (Oatp1, Oatp4, Mrp2, and Ntcp). However, it is unclear whether these changes contribute to reduced RCT, which appears to play a critical role in how TMAO affects CVD (6).

With the caveat that these results are from one laboratory and require independent confirmation, the potential significance of TMAO and the obligate role of the microbiome in TMAO's production can not be under-estimated as a major factor in promoting CVD, especially in precipitating major events associated with CVD.

Medical Relevance

Current recommendations to inhibit the progression of CVD and reduce the risks of major CV events include a healthy diet rich in fruits and vegetables and low in saturated and trans-fats, regular exercise, control of hypertension (if present) by angiotensin-converting enzyme (ACE) inhibitors or angiotensin II receptor blockers (ARBs), and reduction of hypercholesterolemia (if present) by statins.

Recent work suggesting a key role for TMAO in promoting CVD and a critical role for the microbiome in production of TMAO is potentially of great significance. Of particular importance is the potential in using TMAO as a biomarker for CVD risk. Although larger clinical trials are necessary, the size of the reported preliminary trials (>1,000) and the magnitude of the reported risk are encouraging.

For those at risk of CVD, which may include most human beings, control of TMAO levels may prove quite prudent. Of immediate relevance is that supplementation with choline or L-carnitine may be unwise in most circumstances, despite possible benefits in cognitive or muscle function, respectively, unless TMAO levels are carefully monitored. On the other hand, it well known that choline is an important nutrient (11). Low levels of choline can lead to organ dysfunction. Choline is necessary for the production of the neurotransmitter acetylcholine and PC is a critical component of cell membranes. Complete elimination of choline or PC from one's diet would be un-productive as well as difficult to achieve. However excess choline, such as that found in eggs, may be worth avoiding.

Reduction of dietary L-carnitine may be easier to achieve by simple reduction of red meat intake, already a key recommendation for patients with CVD. However, given that carnitine supports skeletal and cardiac muscle function and reports that carnitine supplements can help restore left ventricle ejection fraction during heart failure (12), there may exist special circumstances where L-carnitine supplementation is recommended. Ironically, a meta-analysis of available clinical data suggests that L-carnitine supplementation is helpful after acute MI in reducing mortality and the occurrence of major events (13).

However, it is important to remember that neither choline nor L-carnitine intrinsically have negative physiological effects and indeed may be of clinical benefit. Rather the problem is that some gut flora metabolize them to TMA, which in turn is metabolized into TMAO. Controlling the conversion of choline or L-carnitine into TMAO may prove the more effective strategy. For example, in the case of L-carnitine, a diet low in red meats may be sufficient to suppress TMAO production even with an occasional meal of red meat. One strategy that is not advised is long-term antibiotic use, which proved ineffective in the mouse studies from the Hazen lab. Reduced TMAO levels may be achieved by blocking the biochemical pathways that convert choline or L-carnitine into TMA with, for example, a new small-molecule inhibitor. Alternately, it may be possible to outcompete TMA-producing bacterial species with a pro-biotic containing more benevolent species that lack the ability to metabolize choline or L-carnitine into TMA. Another possibility would be to engineer or discover bacteria that metabolize TMA into a harmless byproduct before it enters the bloodstream. Overall, given that vegans appeared to have the lowest basal levels of TMAO, increased vegetable and fruit intake relative to red meat may be the simplest and safest recommendation.

Conclusions

The pathogenesis of CVD remains a very active topic of research. Recent work showing that the microbiome contributes to the progression of CVD by converting choline and L-carnitine into TMA and eventually TMAO may lead to new therapies for CVD as well as to establish TMAO as an important biomarker. Reduction of L-carnitine and choline in the diet through decreased red meat consumption is presently the simplest and safest means to control TMAO levels. Development of novel therapeutics based on manipulating the microbiome may lead to improved outcomes in CVD.

References

1. Woollard KJ, Geissmann F. Monocytes in atherosclerosis: Subsets and functions. Nat Rev Cardiol 2010;7:77–86.
2. Feldmann R, Geikowski A, Weidner C, Witzke A, Kodelja V, Schwarz T, et al. Foam cell specific LXRa ligand. PloS One 2013;8:e57311.
3. Finn AV, Nakano M, Polavarapu R, Karmali V, Saeed O, Zhao X, et al. Hemoglobin directs macrophage differentiation and prevents foam cell formation in human atherosclerotic plaques. J Am Coll Cardiol 2012;59:166–177.
4. Bennett BJ, de Aguiar Vallim TQ, Wang Z, Shih DM, Meng Y, Gregory J, et al. Trimethylamine-N-oxide, a metabolite associated with atherosclerosis, exhibits complex genetic and dietary regulation. Cell Metab 2013;17:49–60.
5. Tang WHW, Wang Z, Levison BS, Koeth RA, Britt EB, Fu X, et al. Intestinal microbial metabolism of phosphatidylcholine and cardiovascular risk. N Engl J Med 2013;368:1575–1584.
6. Koeth RA, Wang Z, Levison BS, Buffa JA, Org E, Sheehy BT, et al. Intestinal microbiota metabolism of l-carnitine, a nutrient in red meat, promotes atherosclerosis. Nat Med 2013;19:576–585.
7. Wang Z, Klipfell E, Bennett BJ, Koeth R, Levison BS, Dugar B, et al. Gut flora metabolism of phosphatidylcholine promotes cardiovascular disease. Nature 2011;472:57–63.
8. al-Waiz M, Mikov M, Mitchell SC, Smith RL. The exogenous origin of trimethylamine in the mouse. Metabolism 1992;41:135–136.
9. Siri-Tarino PW, Sun Q, Hu FB, Krauss RM. Meta-analysis of prospective cohort studies evaluating the association of saturated fat with cardiovascular disease. Am J Clin Nutr 2010;91:535–546.
10. Ley RE, Hamady M, Lozupone C, Turnbaugh P, Ramey RR, Bircher JS, et al. Evolution of mammals and their gut microbes. Science 2008;320:1647–1651.
11. Zeisel SH, da Costa K-A. Choline: An essential nutrient for public health. Nutr Rev 2009;67:615–623.
12. Matsumoto Y, Sato M, Ohashi H, Araki H, Tadokoro M, Osumi Y, et al. Effects of L-carnitine supplementation on cardiac morbidity in hemodialyzed patients. Am J Nephrol 2000;20:201–207.

13. Dinicolantonio JJ, Lavie CJ, Fares H, Menezes AR, O'Keefe JH. L-Carnitine in the secondary prevention of cardiovascular disease: Systematic review and meta-analysis. Mayo Clin Proc Mayo Clin 2013; available online April 15, 2013.

CHAPTER 10

Trade-offs between Anti-aging Dietary Supplementation and Exercise

In otherwise healthy adults, moderate aerobic exercise extends life span and likely health span by 2–6 years. Exercise improves blood sugar regulation, and resistance exercise increases or maintains muscle mass and is associated with improved cognitive function. On the other hand, evidence for anti-oxidant supplements increasing longevity in humans is lacking. On the contrary, transient hormetic increases in reactive oxygen species (ROS), for example, associated with exercise, are actually associated with increased mammalian health span and life span. Recent studies in humans suggest that anti-oxidants such as vitamins C, E, resveratrol, and acetyl-N-cysteine blunt the beneficial effects of exercise on glucose sensitivity and blood sugar regulation, likely through direct inhibition of ROS signaling. Alternately, other studies suggest that vitamin C has beneficial effects on exercise-associated dysfunction, inhibiting exercise-induced broncho-constriction. These data suggest that there are tradeoffs between potential benefits and harm from anti-oxidant dietary supplementation. Specific biomolecular interactions for each antioxidant also will be important. Omega-3 (n-3) polyunsaturated fatty acids (PUFAs) have anti-inflammatory activity that is not mediated through direct ROS inhibition. Although data are limited in humans, n-3 PUFAs do not seem to blunt blood sugar regulatory benefits of aerobic exercise and actually increase anabolic activity in skeletal muscle. However, another kind of tradeoff may exist with PUFAs, at least for men. A recent large clinical trial demonstrates an association of omega-3 fatty acids blood levels with increased incidence of prostate cancer, especially aggressive prostate cancer. Together these results suggest that there are significant tradeoffs in the use of dietary supplementation for prevention and treatment of diseases associated with aging. Such tradeoffs may result from underlying intertwined homeostatic mechanisms. For most individuals, moderate exercise is of significant benefit. Careful attention to individual and family medical history and personal genomic data may prove essential to make wise dietary and supplement choices to be combined with exercise.

Theories of Aging Have Consequences

There is a cottage industry of aging models in the scientific literature. Like the blind men of legend describing an elephant, each tends to focus on one aspect of aging, often coming to radically different conclusions. Eventually a consensus will be reached in which the correct aspects of each model will be unified and an "elephant" of a theory might emerge. However, until that time, each model, depending on its influence, may have significant consequences for both the direction of scientific research and the development of approaches to slow or even reverse aging.

Perhaps the most influential model for aging research has been Harmon's free-radical theory of aging, which hypothesizes that metabolic activity leads to generation of harmful reactive oxygen species (ROS) that then damage biomolecules, cells, and beyond leading to aging and death (1). This theory was extended to the mitochondrial free-radical theory of aging, which localizes the primary source of damaging free radical production to the mitochondria. Data have accumulated to show that reduction of oxidative stress is associated prolongation of life expectancy in animals, especially invertebrates (for review, see Ristow and Schmeisser (3); see references 133–147 therein). These theories logically led to the idea that anti-oxidants might be able to prolong life span and health span. This idea has become very widespread and popular both in and outside of scientific circles. However, most studies found a lack of correlation between anti-oxidants and human health benefits. Some re ports even suggest that anti-oxidants in humans may promote cancer (for review, see Ristow and Schmeisser (3); see references 163–168 therein) or increased incidence of diseases associated with detrimental effects on longevity (for review, see Ristow and Schmeisser (3); see references 169–175 therein). Remarkably, alternative theories, such as the mitochondrial hormesis or mitohormesis theory (4), propose the opposing hypothesis that appropriate increases in ROS actually promote life span and health span through ROS-based signaling pathways to increase repair, and that increased biomolecular damage observed with aging is due to inadequate ability to respond to endogenous ROS signals. The idea of hormesis in this context is that modest stimulation of ROS is beneficial, but very strong stimulation of ROS after failure to conduct the requested repairs is detrimental (**Figure 10.1**). In fact, caloric restriction, a method of life span enhancement that appears to function from yeast to mice, but may not be effective for primates (5) has been reported at least in invertebrates to transiently increase ROS (6), which is thought to subsequently decrease overall oxidative stress (7) by increasing levels of anti-oxidant defense enzymes. Of course, it is important to keep in mind that ROS-based theories of aging only describe one part of the proverbial elephant.

Anti-Oxidants As Anti-Aging Therapeutics

As mentioned above, no evidence has emerged that anti-oxidants have any benefit for cancer, diabetes, heart disease/atherosclerosis, or high blood pressure (for review, see Ristow and Schmeisser, 3). In fact, anti-oxidants have been associated with a modest increase in some cancers, such as skin cancer in women (8) and interference with radio- and chemotherapies in cancer patients (9).

There is benefit to anti-oxidant therapy in acute situations. For example, n-acetyl cysteine is used to detoxify acetaminophen overdoses or prevent radio-contrast agent damage to the kidney. Benefits of anti-oxidants are likely to depend on the specific antioxidant, as the spectrum of activity will vary greatly by both tissue and intracellular targets. For example, vitamin C appears helpful in diminishing the prevalence and intensity of common colds in subjects undergoing heavy physical activity (10). Among the most well-studied and intriguing anti-oxidants is resveratrol, which is thought to alter metabolism not only by its anti-oxidant properties, but by stimulation of sirtuins, especially by interaction with and activation of SIRT1, which is then thought to activate 5'-adenosine monophosphate-activated protein kinase (AMPK), a kinase involved in integrating and regulating cellular energy levels. Resveratrol has been hypothesized to contribute to the French paradox, i.e., that the deleterious effects of a diet rich in fats may be offset by modest red wine consumption. Horwitz and colleagues identified resveratrol as a sirtuin activator that mimicked calorie restriction and extended life span in yeast, worms, flies, and short-lived fish (11). However, recent data suggest that SIRT1 may not increase longevity in *Caenorhabditis elegans* and *Drosophila melanogaster* (12). In mammals, resveratrol extends life span only in mice fed a fat-rich diet, not normal mice. Although the role of resveratrol to increase life span remains controversial, some evidence suggests that this polyphenol can delay or attenuate many age-associated diseases in animals (for review, see Timmers, 13). Because almost all studies demonstrating efficacy of resveratrol have been performed in rodents, invertebrates, or cell culture (see below), how these effects extend to humans is unclear. To date, only minimal beneficial effects of resveratrol have been directly observed in humans (14).

n-3 PUFA Is a Popular Supplement with Known Anti-Inflammatory Activities

Omega-3 (n-3) polyunsaturated fatty acids (PUFAs) have anti-inflammatory activity that is not mediated through direct inhibition of ROS, but rather via antagonism of n-6 PUFA–induced pro-inflammatory prostaglandin E2, reduced pro-inflammatory master transcription regulator nuclear factor-kB (NF-kB), and increased anti-inflammatory resolvin and protectin levels. Unlike anti-oxidants, mitochondrial peroxidation of n-3 PUFAs leads to production of

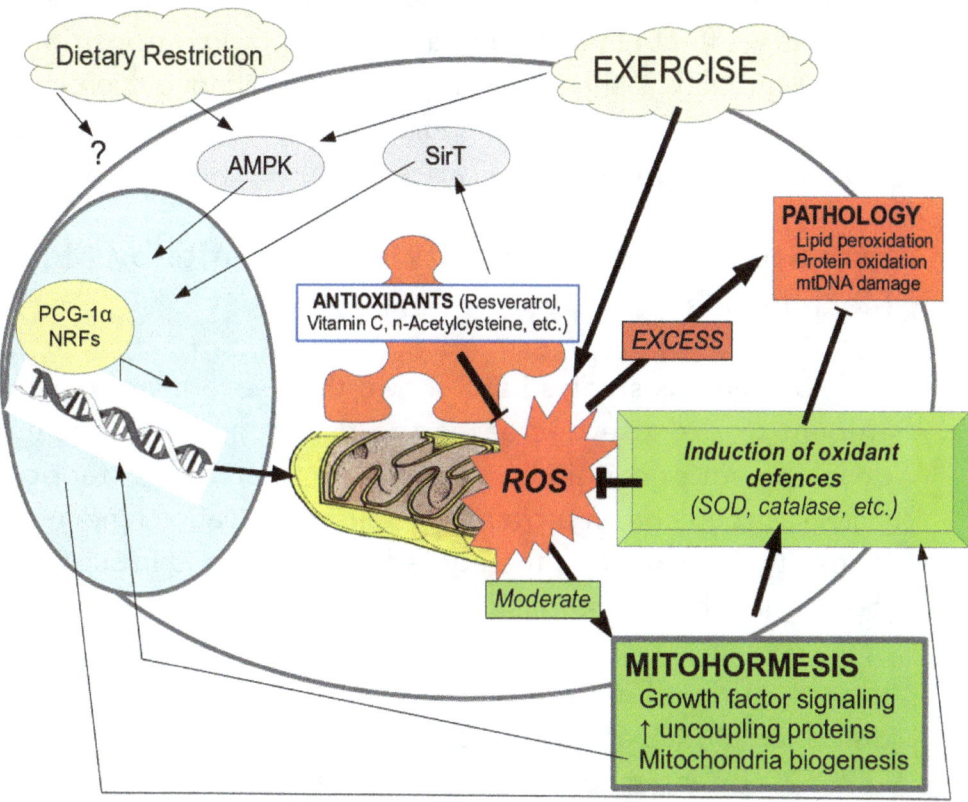

Figure 10.1 Tradeoffs: Anti-oxidants block exercise-induced reactive oxygen species (ROS) stimulation of mitohormesis. Moderate exercise stimulates ROS signaling in the mitochondria, which in turn stimulates mitochondrial biogenesis and anti-oxidant defense biomolecules, resulting in overall reduced oxidative stress. This process has been called mitohormesis, because moderate levels of ROS are needed to produce signaling yet avoid damage associated with excess ROS production. Exercise also signals via 5′-adenosine monophosphate kinase (AMPK) in the cytoplasm to stimulate transcription regulatory factors PCG-1a, NRF1, and NRF2 to increase expression of biomolecules involved in mitochondrial biogenesis and oxidant defenses, such as superoxide dismutases (SODs). Dietary anti-oxidants block exercise-associated induction of ROS, preventing mitohormesis. Both exercise and anti-oxidants separately inhibit excess ROS production that leads to biomoecular damage and pathology, but only exercise alone confers the benefits associated with mitohormesis.

toxic products that may transiently increase ROS, but then indirectly down-regulate ROS by potentiating expression of oxidant defense enzymes such as superoxide dismutase (SOD). n-3 PUFAs may have beneficial effects in arrhythmias, blood lipid composition, and arthritis through its anti-inflammatory actions, although contradictory data in humans have created some controversy (15–17). Interesting anabolic effects on muscle have been observed (see below).

Exercise Extends Longevity and Benefits by Multiple Mechanisms

Numerous studies have shown that performing a minimum amount of exercise decreases the risk of death, prevents the development of certain cancers, lowers the risk of osteoporosis, increases cognition, improves blood sugar regulation, and increases longevity. For example, the age-adjusted increase in survival with jogging in a human study (35-year duration) was 6.2 years in men and 5.6 years in women (18). Exercise also significantly diminished the risk of death in elderly people (19). Relevant to the earlier discussion of ROS and aging, exercise stimulates a modest increase in ROS20–22 and mitochondrial biogenesis (23,24). There are known homeostatic-based tradeoffs of exercise. For example, it has been reported that cardiovascular damage can result from excessive chronic endurance exercise (25). There can always be too much of a good thing. Such trade-offs are universal, potentially quite complex and subtle. Tradeoffs associated with therapeutic intervention may result from underlying the intertwined homeostatic mechanisms, which in turn may result from evolutionary history and tradeoffs to maximize survival.

Clinical Trials Indicate There Are Tradeoffs Between Beneficial and Harmful Effects of Anti-Oxidants and n-3 PUFA Supplementation

Anti-oxidants blunt beneficial effects of exercise.

Recent clinical trials suggest that various anti-oxidants that directly inhibit ROS production ameliorate the benefits of exercise on blood sugar regulation and may blunt other downstream benefits that depend on ROS signaling. Ristow *et al.* (26) subjected 19 untrained and 20 trained, healthy 25- to 35-year-old men to a daily exercise regimen (5 days per week) over 4 weeks. Half of each group received oral antioxidant supplements of 500 mg of vitamin C twice a day and 400 IU of vitamin E once a day. Exercise consisted of 20 min of biking/running, 45 min of circuit training, and 20 min each for warming up and cooling down. In untrained men, overall oxidative stress rose two-fold after 3 days of exercise, as measured by muscle biopsies using a standard redox damage assay measuring thiobarbituric acid reactive substances (TBARS). Anti-oxidant treatment abrogated this response. After 4 weeks, glucose infusion rates (GIR) were measured by a hyper-insulinemic euglycemic clamp, the gold standard for such studies. In both the trained and untrained groups, insulin sensitivity increased significantly.

However, anti-oxidants prevented exercise-dependent induction of insulin sensitivity. There was an additional correlation with increased serum oxidation assayed using TBARS in the untreated group, whereas the vitamin-treated group had TBARS levels similar to baseline. Men who had not received anti-oxidants had increased serum adiponectin, an adipocyte-secreted protein known to be positively correlated with insulin sensitivity in humans and inversely correlated with risk for type 2 diabetes. Adiponectin was unchanged in men receiving anti-oxidants. Similarly fasting insulin levels decreased in untreated men after the exercise regimen, but remained unchanged for the anti-oxidant group. Together these results suggest that anti-oxidants severely reduce the insulin-sensitizing effects of exercise, regardless of whether the subjects were sedentary or already in the habit of regular exercise. At the molecular level, quantitative PCR analysis of muscle biopsies after 4 weeks showed that insulin-sensitivity regulators peroxisome proliferator-activated receptor-c (PPARc) and its two co-activators peroxisome proliferator-activated receptor c co-activator 1 (PGC1a) and PGC1b were strongly increased in the untreated exercise groups, but only marginally increased in the anti-oxidant groups. Expression of PGC1a and PGC1b are known to induce expression of the ROS detoxifying enzymes superoxide dismutase 1 (SOD1), superoxide dismutase 2 (SOD2), and glutathione peroxidase 1 (GPx1). Expression of SOD1, SOD2, and GPx1 were strongly increased only in the exercise groups not receiving anti-oxidants. The authors conclude that exercise induces molecular regulators of insulin sensitivity and anti-oxidant defense mechanisms and that treatment with oral anti-oxidants vitamin C and E abrogate this induction (26). These results are consistent with an earlier study that showed that vitamin C abrogated training-induced mitochondrial biogenesis and increased endurance in humans (27).

In a complementary small study, Petersen *et al.* (28) infused eight healthy young men with n-acetyl cysteine (NAC). NAC is a strong anti-oxidant drug that is approved by the Food and Drug Administration (FDA) for acetaminophen over-dose detoxification and is also used to protect the kidneys from radiocontrast-induced damage and to dissolve mucus by breaking disulfide bonds. NAC is known to quench ROS. Subjects were either infused with saline or NAC (7 days later) and then subjected to exercise tests (running or cycling) lasting at least 45 min until fatigue. Muscle biopsies were analyzed for expression of specific proteins and RNA. NAC blocked the exercise increase in c-Jun N-terminal kinases (JNK) phosphorylation, but not phosphorylation of extracellular signal-regulated kinase 1/2 (ERK1/2) or p38 mitogen-activated protein kinase (MAPK). JNK activation is associated with regulation of genes involved in cell proliferation, apoptosis, inflammation, and DNA repair playing an important role in exercise adaption and is known to be induced by oxidants such as hydrogen peroxide. Phosphorylation of master inflammatory regulator NF-kB was unaffected by exercise, but reduced 14% by NAC. NF-kB may require resistance exercise for induction (29). NAC blocked a threefold increase in expression of manganese SOD, also known as SOD2, similar to the effect seen by Ristow's group. NAC did not block the induction of PGC1a, in contrast to the effects of vitamins C and E observed by Ristow and colleagues. NAC also did not block the induction of heat shock 70 protein (HSP70), interleukin-6 (IL-6) or monocyte chemotactic protein 1 (MCP-1). HSP70 is a chaperone involved in protecting cells from stress. IL-6 is a pro and anti-inflammatory cytokine and a myokine that elevates in response to muscle contraction, involved in the mobilization of extracellular substrates to the muscle. Together these results are consistent with ROS playing a key role in skeletal muscle adaptation to exercise and exogenous anti-oxidants blocking this adaptation.

In an important recent study, resveratrol, an anti-oxidant and sirtuin 1 activator, was found to blunt the beneficial effects of exercise in elderly men, contrary to similar experiments in mice (30). Twenty-seven healthy physically inactive elderly men age (age, 65–1 years; body mass index, 25.4–0.7kg m-2; mean arterial pressure (MAP), 95.8– 2.2 mm Hg; maximal oxygen uptake, 2488 – 72 mL O2 min-1) were divided into two groups: 14 received 250 mg of resveratrol daily, the other 13 received a placebo. The men were subjected to high-intensity training twice a week and full body circuit resistance training once a week for 8 weeks. As expected, exercise resulted in improved oxygen uptake, decreased MAP, increased serum levels of the vasodilator prostacyclin, and decreased plasma levels of blood lipids including low-density lipoprotein and total cholesterol/ high-density lipoprotein ratio and triglyceride concentrations. However, the men in the resveratrol group did not receive these benefits. Protein levels were assessed by either western blot or enzyme-linked immunosorbent assays (ELISAs). Exercise led to a 45% greater (p< 0.05) increase in maximal oxygen uptake in the placebo group than in the resveratrol group. Only the placebo group experienced a decrease in MAP (4.8 – 1.7 mm Hg; p < 0.05). The interstitial level of vasodilator prostacyclin was lower in the resveratrol than in the placebo group after training (980–90 vs. 1174–121 pg mL (1); p< 0.02). Resveratrol also prevented the beneficial effects of exercise on blood lipids. However, vascular cell adhesion protein-1 (VCAM-1) plasma levels, which are considered a marker for progression of atherosclerosis,31 decreased with exercise in both placebo and resveratrol groups, suggesting that resveratrol had no beneficial effect on atherosclerosis, unlike the effect seen in rabbits (32).

Interestingly, these data are in direct contradiction to data from experiments in mice (33–36) and rats (37-38) in which combining exercise with resveratrol acted synergistically to improve exercise capacity and physiological parameters, including blood lipids and cardiovascular risk factors. Increased exercise capacity in rodents has been linked with increased mitochondrial biogenesis induced by SIRT1 activation of AMPK and PGC1a. However, rodents are not appropriate models for human inflammatory diseases (39), and the differential species-specific response to resveratrol may indicate problems with modeling human SIRT1/AMPK and PGC1a physiology in rodents. Consistent with this hypothesis, resveratrol has been shown to reduce (40) or inhibit (41) AMPK and PGC1a in human tissue (**Figure 10.1**).

These experiments expose the necessary caution in determining the potential consequences of earlier human studies that showed ROS and intramuscular free radical-mediated lipid peroxidation to increase with age (42). The obvious conclusion might be to quench ROS with anti-oxidants. However, the studies cited above suggest that this strategy is incorrect. In fact, increased exercise, even in the elderly, improves function. The authors suggest that the level of ROS formation in aged men may not be detrimental to health as previously hypothesized.

Taken together these three studies in humans strongly suggest that anti-oxidant supplementation can blunt the beneficial effects of exercise in humans, likely by abrogating beneficial ROS signaling that stimulates mitochondrial bio- genesis and expression of oxidant defense biomolecules (**Figure 10.1**). In other words, combining anti-oxidant supplementation with exercise blocks the beneficial mitohormesis associated with exercise. More work will be necessary to ascertain whether the differences in the spectrum of changes seen between theses studies was due to differences in response to the specific anti-oxidants, different experimental protocols or other parameters such as the age of the subjects. Although anti-oxidants may block the positive effects of exercise, there are scenarios where blocking ROS may be helpful in ameliorating the negative effects of exercise.

Vitamin C ameliorates exercise-induced broncho-constriction

As expected for a system built by Darwinian evolution, there are no universally beneficial activities. There are common detrimental effects of exercise. For example, a transient narrowing of the airways in the lung that occurs during or after exercise is called exercise-induced bronchoconstriction (EIB). EIB involves a decrease of 10% or greater in forced expiratory volume in 1 sec (FEV1). EIB occurs in about 10% of the general population and up to 50% in participants of some sports. The underlying physiological causes of EIB are not well understood, but may involve respiratory water loss, which leads to release of pro-inflammatory molecules such as histamine, leukotrienes, and prostaglandins. Increased exhaled nitric oxide may also be involved. It is hypothesized that vitamin C antagonizes prostaglandin metabolism, and 2-week administration of vitamin C was observed to reduce post-exercise bronchoconstrictors LTC4-LTE4 and PGD2.

A new meta-analysis (43) of three placebo controlled trials (44–46) that studied the effects of vitamin C on EIB suggests that there is an 8.4% decrease in the pooled effect for post-exercise FEV1 when vitamin C is given before exercise and a 48% decrease in the relative pooled effect (43). Given the small pooled sample size of 40 and an earlier flawed meta-analysis that did not find a significant effect of vitamin C, it would be quite useful for follow-up studies to confirm the conclusion of this meta-analysis, especially given the potential for cherry-picking studies for statistical analysis. Analysis of the specific molecular mechanisms involved is also warranted, because it would be useful to prove that the anti-oxidant capability of vitamin C is important to its effect on EIB. However, the conclusions of this study dovetail nicely with data suggesting that vitamin C lowers the incidence and severity of the common cold in people under heavy acute physical stress (10,47). Trials to test the effects of other anti-oxidants to control the severity of EIB and the common cold may also be warranted.

The anti-inflammatory n-3 PUFA augments muscle anabolism after meals or strength training

Two recent small clinical studies showed that n-3 PUFA supplementation increases skeletal muscle protein synthesis in humans under a hyperaminoacidemic–hyperinsulinemic clamp. In this protocol, insulin and amino acids are infused over several hours to simulate the effects of a meal in both young and old subjects (48,49) In a third clinical study, fish oil supplementation rich in n-3 PUFA augmented the results of strength training in elderly women (50). Sedentary young (48) or elderly (49) subjects were given stable isotope phenylalanine tracers to study muscle protein synthesis over a 3-hr hyperaminoacidemic–hyperinsulinemic clamp following 8 weeks of daily ingestion of 4 grams of n-3 PUFA. Incorporation of the tracer into muscle protein was used to assess muscle protein synthesis (MPS) rates. In both studies, basal MPS did not change with n-3 PUFA supplementation. However, the hyperaminoacidemic–

hyperinsulinemic clamp method showed an increase in MPS following supplementation with addition to n-3 PUFA. In both studies, the amount of signaling in the mammalian target of rapamycin (mTOR)/p70 growth regulatory signaling pathway increased by about 50% as assessed by increases in phospho-mTOR-Ser2448 and p70s6kThr389. This study suggests that n-3 PUFAs augment MPS in the presence of high levels of insulin and amino acids, which would typically occur after meals, and that n-3 PUFAs increase the anabolic rate in skeletal muscle.

In a study of 45 elderly women (*64 years old) divided into three groups of 15 subjects who received either strength training for 90 days, strength training and 2 grams/day of fish oil for 90 days or 150 days, all groups showed increased peak torque and torque development for the muscles tested (knee flexor and extensor, plantar, and dorsiflexor) and chair rising performance after training. However, the effect was greater among the groups receiving fish oil (50). The authors conclude that fish oil supplementation contributes to improved muscle strength and functional capacity.

Together these data suggest that n-3 PUFAs act to increase the anabolic rate (synthesis of muscle protein) in both sedentary and active individuals. So the question arises: Are there any obvious tradeoffs?

n-3 PUFAs are associated with increased risk of prostate cancer

A very large case cohort study was recently concluded on the subjects of the selenium and vitamin E cancer prevention trial (51). In all, 834 men diagnosed with prostate cancer were designated as the case subjects; 156 of these had high-grade prostate cancer. The sub-cohort included 1393 men of similar age and race.

Compared with men in the lowest quartiles of n-3 PUFA, men in the highest quartile were at increased risk for low-grade (hazard ratio [HR] = 1.44, 95% confidence interval [CI] = 1.08–1.93), high-grade (HR = 1.71, 95% CI = 1.00–2.94), and total prostate cancer (HR = 1.43, 95% CI = 1.09–1.88). Similar associations were observed for individual long-chain n-3 PUFAs. Higher linoleic acid (x-6) was associated with decreased risks of low-grade (HR = 0.75, 95% CI = 0.56–0.99) and total prostate cancer (HR = 0.77, 95% CI = 0.59–1.01). However, no dose response was observed (51).

This study agrees with previous reports of increased prostate cancer risk among men with high blood concentrations of n-3 PUFAs in three out of four studies published after 2007 (52–54). The consistency of these findings implicates these fatty acids in prostate tumorigenesis. The authors suggest that men should consider the potential risk of prostate cancer associated with n-3 PUFA supplementation.

It is intriguing that n-3 PUFAs act to both stimulate anabolism of muscle and contribute an increased risk of prostate cancer. We recommend testing the hypothesis that n-3 PUFAs act indirectly on or most likely downstream of the androgen receptor, which is a known target for muscle enhancement by agonists and prostate cancer therapy by antagonists. In terms of anti-oxidant activity, there are data to support the idea that n-3 PUFAs are potential pro-oxidants that may stimulate ROS by peroxidation after incorporation into cell membranes (for review, see al Gubory, 55), which may also be another mechanism by which they promote prostate cancer.

Medical Implications

Significant tradeoffs are associated with consumption of anti-oxidant supplements to promote health span and life span. Given reasonable mechanisms of action and similar results obtained using several different anti-oxidants argue for caution. However, given the small size of studies on antioxidants and exercise, larger studies in diverse populations are warranted. Also, other commonly used anti-oxidants should be investigated for their ability to block the benefits of physical activity, to investigate how general the effect is, if there are exceptions, and to ensure that informed decisions are possible.

The potential disadvantages of anti-oxidant and n-3 PUFA supplementation suggest that their use as supplements be evaluated on an individual basis. A person's age, sex, personal and family medical history, and personal genomics, if available, might be used to determine if and how much dietary supplementation is wise. For example, a man with a family history of prostate cancer or increased risk due to lifestyle or genomic variations may wish to avoid n-3 PUFA supplementation, whereas such precautions obviously would not apply to women. In fact, a recent study suggests that n-3 PUFAs reduce the risk of arthritis in women (56), suggesting that n-3 PUFA supplementation or a diet rich in fish may be beneficial in women. The most difficult clinical situations arise when two conditions conflict. For example, a pre-diabetic patient suffering from asthma, who is routinely engaging in aerobic exercise would have a difficult decision to make concerning whether they take vitamin C to reduce their risk of bronchial constriction-asthma attack during and after exercise. Common sense dictates that the relative severity of each potential problem be considered to make a wise decision.

Appropriate timing may be a critical way to intelligent use of anti-oxidant supplements. Treatment to ameliorate acute symptoms is one scenario. For example, runners who suffer from exercise-induced bronchial constriction might optimize timing for consumption of vitamin C so as to reduce the possibility of an attack, but still benefit from the exercise. Perhaps an appropriate pro-oxidant can be used subsequently to neutralize vitamin C's anti-oxidant effect, after preventing the potential breathing problem. Another similar scenario would be early treatment with zinc and vitamin C supplements to reduce the severity of the common cold (57).

Given substantial life span and health span benefits, moderate exercise is to be encouraged for all of those where such activity is not contraindicated due to pre-existing medical conditions. For many, perhaps dietary supplementation should be made on the basis of need, such as increasing low vitamin D3 or B12 levels in the elderly. For most the best way to maximize anti-oxidant and n-3 PUFA intake may be a balanced diet rich in fruits, vegetables and fish.

Conclusions

Potent anti-oxidant supplements blunt the beneficial effects of exercise, the only activity that has been associated with increased longevity in humans. On the other hand, there are beneficial effects of anti-oxidants, for example, vitamin C blunts bronchoconstriction after intense exercise. n-3 PUFAs have anti-inflammatory activity affecting multiple organs and stimulate skeletal muscle growth. Unfortunately, these fatty acids recently have been associated with an increased incidence of prostate cancer. These results suggest that there are significant tradeoffs in the use of dietary supplementation to prevent and treat diseases associated with aging. Such tradeoffs may result from the underlying, intertwined homeostatic mechanisms, which in turn may result from evolutionary history and tradeoffs therein. We hypothesize that many such tradeoffs remain to be elucidated. For most individuals, moderate exercise is clearly of significant benefit.

References

1. Harman D. Aging: A theory based on free radical and radiation chemistry. J Gerontol 1956;11:298–300.
2. Harman D. The biologic clock: The mitochondria? J Am Geriatr Soc 1972;20:145–147.
3. Ristow M, Schmeisser S. Extending life span by increasing oxidative stress. Free Radic Biol Med 2011;51:327–336.
4. Tapia PC. Sublethal mitochondrial stress with an attendant stoichiometric

augmentation of reactive oxygen species may precipitate many of the beneficial alterations in cellular physiology produced by caloric restriction, intermittent fasting, exercise and dietary phytonutrients: "Mitohormesis" for health and vitality. Med Hypotheses 2006;66:832–843.

5. Mendelsohn AR, Larrick JW. Dietary restriction: Critical cofactors to separate health span from life span benefits. Rejuvenation Res 2012;15:523–529.

6. Schulz TJ, Zarse K, Voigt A, Urban N, Birringer M, Ristow M. Glucose restriction extends Caenorhabditis elegans life span by inducing mitochondrial respiration and increasing oxidative stress. Cell Metab 2007;6:280–293.

7. Buchowski MS, Hongu N, Acra S, Wang L, Warolin J, Roberts LJ. Effect of modest caloric restriction on oxidative stress in women, a randomized trial. PLoS One 2012;7.

8. Hercberg S, Ezzedine K, Guinot C, Preziosi P, Galan P, Bertrais S, et al. Antioxidant supplementation increases the risk of skin cancers in women but not in men. J Nutr 2007;137:2098–2105.

9. Bjelakovic G, Nikolova D, Gluud LL, Simonetti RG, Gluud C. Mortality in randomized trials of antioxidant supplements for primary and secondary prevention: Systematic review and meta-analysis. JAMA 2007;297:842–857.

10. Hemila H, Chalker E. Vitamin C for preventing and treating the common cold. Cochrane Database Syst Rev 2013;1:CD000980.

11. Mohar DS, Malik S. The sirtuin system: The holy grail of resveratrol? J Clin Exp Cardiol 2012;3.

12. Burnett C, Valentini S, Cabreiro F, Goss M, Somogyvari M, Piper MD, et al. Absence of effects of Sir2 overexpression on lifespan in *C. elegans* and Drosophila. Nature 2011;477:482–485.

13. Timmers S, Auwerx J, Schrauwen P. The journey of resveratrol from yeast to human. Aging 2012;4:146–158.

14. Cottart C-H, Nivet-Antoine V, Beaudeux J-L. Review of re- cent data on the metabolism, biological effects, and toxicity of resveratrol in humans. Mol Nutr Food Res 2013; published online June 6, 2013, doi: 10(1)002/mnfr.201200589.

15. Deckelbaum RJ, Torrejon C. The omega-3 fatty acid nutritional landscape: Health benefits and sources. J Nutr 2012;142:587S–591S.

16. Borghi C, Pareo I. Omega-3 in antiarrhythmic therapy: Cons position. High Blood Press Cardiovasc Prev Off J Ital Soc Hypertens 2012;19:207–211.

17. Trimarco B. Omega-3 in antiarrhythmic therapy: Pros position. High Blood Press Cardiovasc Prev Off J Ital Soc Hypertens 2012;19:201–205.

18. Schnohr P, Marott JL, Lange P, Jensen GB. Longevity in male and female joggers: The Copenhagen City Heart Study. Am J Epidemiol 2013;177:683–689.

19. Buchman AS, Yu L, Boyle PA, Shah RC, Bennett DA. Total daily physical activity and longevity in old age. Arch Intern Med 2012;172:444–446.

20. Davies KJA, Quintanilha AT, Brooks GA, Packer L. Free radicals and tissue damage produced by exercise. Biochem Biophys Res Commun 1982;107:1198–

1205.
21. Chevion S, Moran DS, Heled Y, Shani Y, Regev G, Abbou B, *et al*. Plasma antioxidant status and cell injury after severe physical exercise. Proc Natl Acad Sci USA 2003;100:5119–5123.
22. Powers RW, Kaeberlein M, Caldwell SD, Kennedy BK, Fields S. Extension of chronological life span in yeast by decreased TOR pathway signaling. Genes Dev 2006;20:174–184.
23. Konopka AR, Douglass MD, Kaminsky LA, Jemiolo B, Trappe TA, Trappe S, *et al*. Molecular adaptations to aerobic exercise training in skeletal muscle of older women. J Gerontol A Biol Sci Med Sci 2010;65A:1201–1207.
24. Konopka AR, Suer MK, Wolff CA, Harber MP. Markers of human skeletal muscle mitochondrial biogenesis and quality control: Effects of age and aerobic exercise training. J Gerontol A Biol Sci Med Sci 2013; published online July 20, 2013, doi: 10(1)093/gerona/glt107.
25. Patil HR, O'Keefe JH, Lavie CJ, Magalski A, Vogel RA, McCullough PA. Cardiovascular damage resulting from chronic excessive endurance exercise. Mol Med 2012;109: 312–321.
26. Ristow M, Zarse K, Oberbach A, Kloting N, Birringer M, Kiehntopf M, *et al*. Antioxidants prevent health-promoting effects of physical exercise in humans. Proc Natl Acad Sci USA 2009;106:8665–8670.
27. Gomez-Cabrera M-C, Domenech E, Romagnoli M, Arduini A, Borras C, Pallardo FV, *et al*. Oral administration of vitamin C decreases muscle mitochondrial biogenesis and hampers training-induced adaptations in endurance performance. Am J Clin Nutr 2008;87:142–149.
28. Petersen AC, McKenna MJ, Medved I, Murphy KT, Brown MJ, Della Gatta P, *et al*. Infusion with the antioxidant N- acetylcysteine attenuates early adaptive responses to exercise in human skeletal muscle: NAC and exercise-induced cell signalling. Acta Physiol 2012;204:382–392.
29. Jimenez-Jimenez R, Cuevas MJ, Almar M, Lima E, Garcia-Lopez D, De Paz JA, *et al*. Eccentric training impairs NF-kappaB activation and over-expression of inflammation-related genes induced by acute eccentric exercise in the elderly. Mech Ageing Dev 2008;129:313–321.
30. Gliemann L, Schmidt JF, Olesen J, Biensø RS, Peronard SL, Grandjean SU, *et al*. Resveratrol blunts the positive effects of exercise training on cardiovascular health in aged men. J Physiol 2013; published online August 16, 2013, doi: 10(1)113/jphysiol.2013.258061
31. Peter K, Nawroth P, Conradt C, Nordt T, Weiss T, Boehme M, *et al*. Circulating vascular cell adhesion molecule-1 correlates with the extent of human atherosclerosis in contrast to circulating intercellular adhesion molecule-1, e-selectin, p-selectin, and thrombomodulin. Arterioscler Thromb Vasc Biol 1997;17:505–512.
32. Matos RS, Baroncini LAV, Pre coma LB, Winter G, Lambach PH, Caron EY, *et*

al. Resveratrol causes antiatherogenic effects in an animal model of atherosclerosis. Arq Bras Cardiol 2012;98:136–142.
33. Menzies KJ, Singh K, Saleem A, Hood DA. Sirtuin 1-mediated effects of exercise and resveratrol on mitochondrial biogenesis. J Biol Chem 2013;288:6968–6979.
34. Murase T, Haramizu S, Ota N, Hase T. Suppression of the aging-associated decline in physical performance by a combination of resveratrol intake and habitual exercise in senescence-accelerated mice. Biogerontology 2009;10:423–434.
35. Price NL, Gomes AP, Ling AJY, Duarte FV, Martin-Montalvo A, North BJ, *et al.* SIRT1 is required for AMPK activation and the beneficial effects of resveratrol on mitochondrial function. Cell Metab 2012;15:675–690.
36. Lagouge M, Argmann C, Gerhart-Hines Z, Meziane H, Lerin C, Daussin F, *et al.* Resveratrol improves mitochondrial function and protects against metabolic disease by activating SIRT1 and PGC-1alpha. Cell 2006;127:1109–1122.
37. Dolinsky VW, Jones KE, Sidhu RS, Haykowsky M, Czubryt MP, Gordon T, *et al.* Improvements in skeletal muscle strength and cardiac function induced by resveratrol during exercise training contribute to enhanced exercise performance in rats. J Physiol 2012;590:2783–2799.
38. Hart N, Sarga L, Csende Z, Koltai E, Koch LG, Britton SL, *et al.* Resveratrol enhances exercise training responses in rats selectively bred for high running performance. Food Chem Toxicol Int J Publ Br Ind Biol Res Assoc 2013; online February 17, 2013.
39. Seok J, Warren HS, Cuenca AG, Mindrinos MN, Baker HV, Xu W, *et al.* Genomic responses in mouse models poorly mimic human inflammatory diseases. Proc Natl Acad Sci USA 2013;110:3507–3512.
40. Yoshino J, Conte C, Fontana L, Mittendorfer B, Imai S, Schechtman KB, *et al.* Resveratrol supplementation does not improve metabolic function in non-obese women with normal glucose tolerance. Cell Metab 2012;16:658–664.
41. Skrobuk P, von Kraemer S, Semenova MM, Zitting A, Koistinen HA. Acute exposure to resveratrol inhibits AMPK activity in human skeletal muscle cells. Diabetologia 2012;55: 3051–3060.
42. Bailey DM, McEneny J, Mathieu-Costello O, Henry RR, James PE, McCord JM, *et al.* Sedentary aging increases resting and exercise-induced intramuscular free radical formation. J Appl Physiol 2010;109:449–456.
43. Hemila H. Vitamin C may alleviate exercise-induced bronchoconstriction: A meta-analysis. BMJ Open 2013;3: e002416.
44. Schachter EN, Schlesinger A. The attenuation of exercise-induced bronchospasm by ascorbic acid. Ann Allergy 1982; 49:146–151.
45. Cohen HA, Neuman I, Nahum H. Blocking effect of vitamin C in exercise-induced asthma. Arch Pediatr Adolesc Med 1997;151:367–370.
46. Tecklenburg SL, Mickleborough TD, Fly AD, Bai Y, Stager JM. Ascorbic acid

supplementation attenuates exercise-induced bronchoconstriction in patients with asthma. Respir Med 2007;101:1770–1778.

47. Hemila H. Vitamin C and common cold incidence: A review of studies with subjects under heavy physical stress. Int J Sports Med 1996;17:379–383.

48. Smith GI, Atherton P, Reeds DN, Mohammed BS, Rankin D, Rennie MJ, et al. Omega-3 polyunsaturated fatty acids augment the muscle protein anabolic response to hyperinsulinaemia–hyperaminoacidaemia in healthy young and middle-aged men and women. Clin Sci 2011;121:267–78.

49. Smith GI, Atherton P, Reeds DN, Mohammed BS, Rankin D, Rennie MJ, et al. Dietary omega-3 fatty acid supplementation increases the rate of muscle protein synthesis in older adults: A randomized controlled trial. Am J Clin Nutr 2011;93:402–412.

50. Rodacki CL, Rodacki AL, Pereira G, Naliwaiko K, Coelho I, Pequito D, et al. Fish-oil supplementation enhances the effects of strength training in elderly women. Am J Clin Nutr 2012;95:428–436.

51. Brasky TM, Darke AK, Song X, Tangen CM, Goodman PJ, Thompson IM, et al. Plasma phospholipid fatty acids and prostate cancer risk in the SELECT Trial. J Natl Cancer Inst 2013;105:1132–1141.

52. Brasky TM, Till C, White E, Neuhouser ML, Song X, Goodman P, et al. Serum phospholipid fatty acids and prostate cancer risk: Results from the Prostate Cancer Prevention Trial. Am J Epidemiol 2011;173:1429–1439.

53. Park S-Y, Wilkens LR, Henning SM, Le Marchand L, Gao K, Goodman MT, et al. Circulating fatty acids and prostate cancer risk in a nested case-control study: The Multiethnic Cohort. Cancer Causes Control CCC 2009;20:211–223.

54. Crowe FL, Allen NE, Appleby PN, Overvad K, Aardestrup IV, Johnsen NF, et al. Fatty acid composition of plasma phospholipids and risk of prostate cancer in a case-control analysis nested within the European Prospective Investigation into Cancer and Nutrition. Am J Clin Nutr 2008;88: 1353–1363.

55. Al-Gubory KH. Mitochondria: Omega-3 in the route of mitochondrial reactive oxygen species. Int J Biochem Cell Biol 2012;44:1569–1573.

56. Di Giuseppe D, Wallin A, Bottai M, Askling J, Wolk A. Long-term intake of dietary long-chain n-3 polyunsaturated fatty acids and risk of rheumatoid arthritis: A prospective cohort study of women. Ann Rheum Dis 2013; doi:10(1)136/annrheumdis-2013-203338.

57. Maggini S, Beveridge S, Suter M. A combination of high-dose vitamin C plus zinc for the common cold. J Int Med Res 2012;40:28–42.

Chapter 11

Sleep facilitates clearance of metabolites from the brain: Glymphatic function in aging and neurodegenerative diseases

Decline of cognition and increasing risk of neurodegenerative diseases are major problems associated with aging in humans. Of particular importance is how the brain removes potentially toxic biomolecules that accumulate with normal neuronal function. Recently a biomolecule clearance system using convective flow between the cerebrospinal fluid (CSF) and interstitial fluid (ISF) to remove toxic metabolites in the brain was described. Xie and colleagues now report that in mice clearance activity of this so-called "glymphatic system" is strongly stimulated by sleep and is associated with an increase in interstitial volume, possibly by shrinkage of astroglial cells. Moreover, anesthesia and attenuation of adrenergic signaling can activate the glymphatic system to clear potentially toxic proteins known to contribute to the pathology of Alzheimer's disease (AD) such as amyloid beta (aBeta). Clearance during sleep is as much as two-fold faster than during waking hours. These results support a new hypothesis to answer the age-old question of why sleep is necessary. Glymphatic dysfunction may pay a hitherto unsuspected role in the pathogenesis of neurodegenerative diseases as well as maintenance of cognition. Furthermore, clinical studies suggest that quality and duration of sleep may be predictive of the onset of AD, and that quality sleep may significantly reduce the risk of AD for Apolipoprotein E (ApoE) e4 carriers, who have significantly greater chances of developing AD. Further characterization of the glymphatic system in humans may lead to new therapies and methods of prevention of neurodegenerative diseases. A public health initiative to ensure adequate sleep among middle-aged and older people may prove useful in preventing AD, especially in ApoE e4 carriers.

Introduction

Decline of cognition and increasing risk of neurodegenerative diseases are major problems associated with aging in humans. For example, after age 85, the risk of Alzheimer's disease is 50% (1). Understanding the pathology of neuronal loss and dysfunction with aging is critical to developing therapies or preventative measures. Of particular importance is how the brain removes potentially toxic biomolecules that accumulate with normal neuronal function.

Of special interest is the metabolism of the beta amyloid protein. Abeta has numerous normal functions, as a transcription factor, activation of kinases, oxidative stress protection, cholesterol transport and protection from microbes. Unfortunately, Abeta can damage neurons and is the key component of the amyloid plaques found in Alzheimer's Disease. Known mechanisms of Abeta clearance include uptake by microglial phagocytosis, receptor-mediated transport across the blood vessel walls and degradation by enzymes: glutamate carboxypeptidase II, insulin-degrading enzyme, matrix metalloproteinase-9, neprilysin, and others (2). Similar mechanisms exist for other toxic, potentially pathological brain proteins such as alpha-synuclein and tau. For example, tau is degraded by autophagic and proteasomal degradative systems and a variety of proteases which include aminopeptidases, HTRA1, calpain, caspases, and thrombin (3).

The recent discovery of the glymphatic clearance system provides fresh insight into how the brain clears potentially toxic metabolites (4) (5) (6). Unlike most of the other organs in the body, the brain lacks lymphatic vasculature for clearing excessive interstitial fluid, metabolic waste products and unneeded extracellular biomolecules, such as Abeta (**Figure 11.1**). However, the brain does have the glymphatic system which accomplishes the same purpose and protects fragile neurons from the toxic side-effects of metabolism.

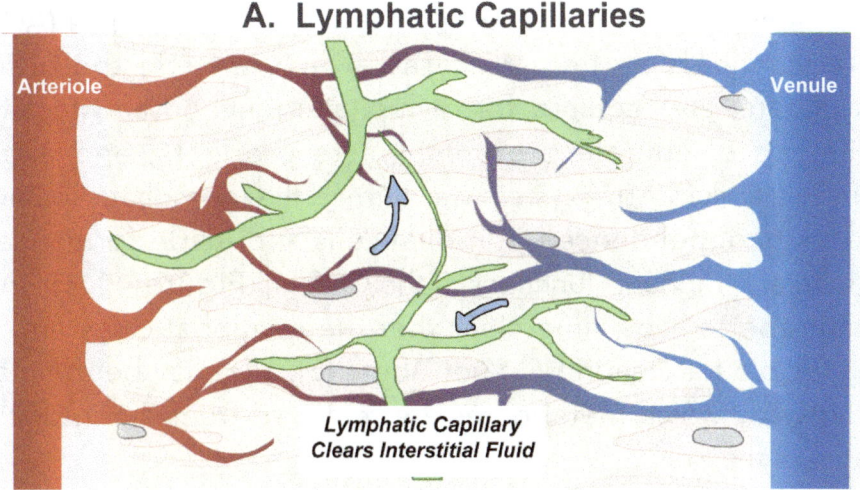

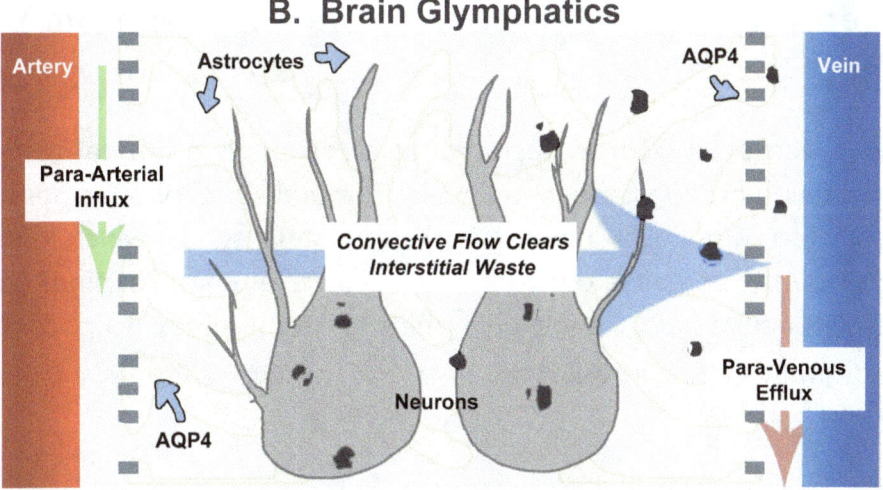

Figure 11.1   Glymphatic system works analogously to the lymphatic system. Lymphatic vessels clears interstitial fluid of waste (A). In the glymphatic system, convective flow between cerebrospinal fluid entering through arteries and interstitial fluid exiting through veins removes biomolecular waste in the brain.

The glymphatic system uses convective flow to clear biomolecules. Cerebrospinal fluid (CSF) enters the brain parenchyma (functional parts) along a para-arterial route and exchanges with the interstitial fluid (ISF). The ISF carries extracellular solutes from the interstitial (extracellular) space in the brain along para-venous drainage pathways. Aquaporin 4 (AQP4) water channels in the astrocytes, which encircle the brain's vasculature, are required for glymphatic function and facilitate clearance of soluble proteins, waste products, and excess extracellular fluid. This system was termed the 'glymphatic' pathway due to its dependence upon glial cells and its performance of peripheral 'lymphatic' functions in the CNS [4] (**Figure 11.1**). From initial reports it became clear that the glymphatic system was responsible for clearing 65% of Abeta [5], and may be the major pathway by which metabolites are removed from the interstitial spaces of the brain.

Sleep facilitates clearance of metabolites from the brain

Xie et al (7) now report that sleep plays a critical role in maintaining homeostasis in the brain. Natural sleep or anesthesia is associated with a 60% increase in the interstitial space, which in turn significantly increases convective exchange of cerebrospinal fluid with interstitial fluid via the glymphatic system, removing potential neurotoxic waste products such as Abeta that accumulate during waking periods.

This mechanism contributing to the restorative nature of sleep was unknown in spite of the numerous negative consequences of sleep deprivation: prolonged reaction times, reduced cognitive performance (8), increased risk of seizures or hallucinations (9), (10) and even death in a matter of days for flies [11] or weeks for rodents (12). In humans, the syndrome known as progressive familial insomnia leads to dementia and death (13).

The recent discovery of the glymphatic system prompted the authors to test the hypothesis that sleep might prevent the negative consequences of insomnia by altering glymphatic function. Using sophisticated neuroscience and cell biology techniques, Xie *et al.* discovered that infusion of a small (3 kD) green fluorescent tracer molecule (FITC-dextran) via CSF into brains during the day, when the mice were asleep, rapidly penetrated the brain (the specific regions noted were the periarterial spaces, the subpial regions, and the brain parenchyma). The tracer was visualized by two-photon imaging, in which two low-energy infrared photons need to simultaneously be absorbed by the tracer to stimulate emission, providing very high spacial resolution. Electrocortigraphy (EcoG), in which electrical signals from electrodes physically localized within the brain are analyzed, was used to determine that the mice were indeed asleep by the tell-tale appearance of slow wave patterns. After waking the mice by gently touching their tails, CSF loaded with a second fluorescent tracer of a different color, Texas red-dextran (3kD) was injected. This time, tracer influx was significantly reduced. EcoG showed decreased slow-wave activity, consistent with awake animals. To confirm these results, the experiments were repeated in the evening when the animals are awake. This time the first tracer, FITC-dextran in CSF was barely detectable within the brain. After 30 minutes, the mice were anesthetized with ketamine/xylene, and the second tracer Texas red/dextran was injected. This tracer immediately entered the brain, finding its way to the same regions previously observed in the earlier experiment when the mice were sleeping. The authors concluded that the brain is more permeable to CSF when the animals are asleep (7).

How does this occur?

There are at least two possible mechanisms to explain the restriction of CSF flow within the awake brain. Perhaps arterial pulse waves that force the CSF inward are greater during wakefulness, or the interstitial space is reduced, increasing resistance to convective flow. Given that arterial blood pressure is higher during physical activity, and high pressure would presumably drive the CSF out into the interstitial spaces, the first hypothesis was deemed unlikely. So the authors investigated the second hypothesis. To measure the brain interstitial space they employed the widely used real-time iontophoretic tetramethylammonium (TMA) method in head-fixed mice. In the TMA method, an electrode with two components is embedded into the mouse brain. The first component uses an electric pulse to release a small amount of positively charged TMA, to diffuse into the IFS. The second component is positioned a defined distance away from the first and specifically detects the presence of TMA. The amount of TMA detected in a defined time is proportional to the volume of the IFS.

TMA recordings when the mice were asleep indicated that the IFS volume was 23.4 +/- 1.9%, compared to 14.1% +/- 1.8% when the mice were awake in the evening. EcoG readings confirmed the state of consciousness of the animals was as expected. IFS volume was increased when the animals were asleep, confirming the hypothesis. Furthermore, in experiments in which the waking mice were anesthetized, IFS volume consistently increased >60% from 13.6 +/- 1.6% to 22.7 +/- 1.3%. Cortical IFS varies from 13-15% when mice are awake to 22 to 24% when they are asleep. Similar results were found when measurements were made 150 or 300 uM below the brain's surface. In addition the tortuosity (twistedness) of the IFS was unchanged (7). <ins>The increased IFS volume during sleep decreases tissue resistance to IFS flux and</ins>

inward movement of CSF (**Figure 11.2**).

The next question is whether the increased convection during sleep results in increased clearance of metabolites or potentially neurotoxic biomolecules. Since the glymphatic system had been previously reported to clear 65% of Abeta, Abeta clearance was measured during sleep, wakefulness, and under anesthesia in mice using radio labeled I^{125}-Abeta. Brains were isolated 10-240 minutes later and Abeta retention was measured. Consistent with the important role of the glymphatic system in clearance, Abeta was cleared twice as fast in sleeping and anesthetized mice as in awake mice. To completely rule out other transport effects (e.g. Abeta is also removed by receptor-mediated transport across the blood-brain-barrier (BBB)), radiolabeled inulin was also observed to be cleared twice as fast in sleeping and anesthetized mice (7).

What triggers the sleep-associated increase in IFS and biomolecule clearance? Circadian rhythms are not a likely explanation, because anesthesia can be administered at any time to reduce interstitial space and activate clearance. Instead, the authors hypothesized that state of arousal itself was responsible. In particular, locus coruleus noradrenergic signaling, which is associated with alertness/wakefulness, may be the direct cause. In other words, adrenergic signaling in the awake state reduces interstitial volume and decreases clearance. To test the hypothesis, awake mice were treated with a cocktail of adrenergic antagonists 15 minutes before CSF loaded with tracer was administered. As expected, tracer influx was increased in the presence of the antagonists. Next Xie et al measured the effects of adrenergic inhibition on interstitial space using the TMA method after directly applying the inhibitors to the exposed cortical surface of the brains of awake mice. Interstitial volume increased from 14.3% +/- 5.2% to 22.6 +/- 1.2%., similar to that seen with sleep or anesthesia.

Consistent with the known effects blocking adrenergic signaling, more slow wave patterns were observed with EcoG measurements indicating a more relaxed, sleep-like state was induced (7). Therefore, the adrenergic system not only regulates wakefulness, but also interstitial volume and glymphatic clearance.

Which cells are involved in mediating interstitial clearance? That remains unknown, although sleep-mediated shrinkage of the astroglial and other neuronal support cells would be the leading hypothesis.

Overall, Xie *et al.* provide evidence that a key purpose of sleep as a physiological process is to clear potentially toxic waste products of metabolism, such as Abeta, which is known to trigger irreversible neuronal injury, interfere with synaptic transmission and is the critical component of toxic amyloid deposits in Alzheimer's disease.

We can speculate that the accumulation of waste products somehow helps triggers sleep in conjunction with circadian rhythms and that increased interstitial volume associated with sleep may be involved in long-term memory consolidation, which occurs during sleep. These ideas might be the subject of subsequent work.

This study is a potential breakthrough in our under-standing of sleep and brain physiology and has significant implications. However, to date direct evidence linking the glymphatic system to a critical role in sleep has been reported by primarily by one group. This work needs to be independently replicated, especially in other mammals, including humans. So far, evidence for the existence of the glymphatic system has been reported by the same group for rats [6]. Unfortunately, direct access to the human brain in more problematic than in mice, and such work may require modification of

Chapter 11   Sleep clears metabolites from the brain   183

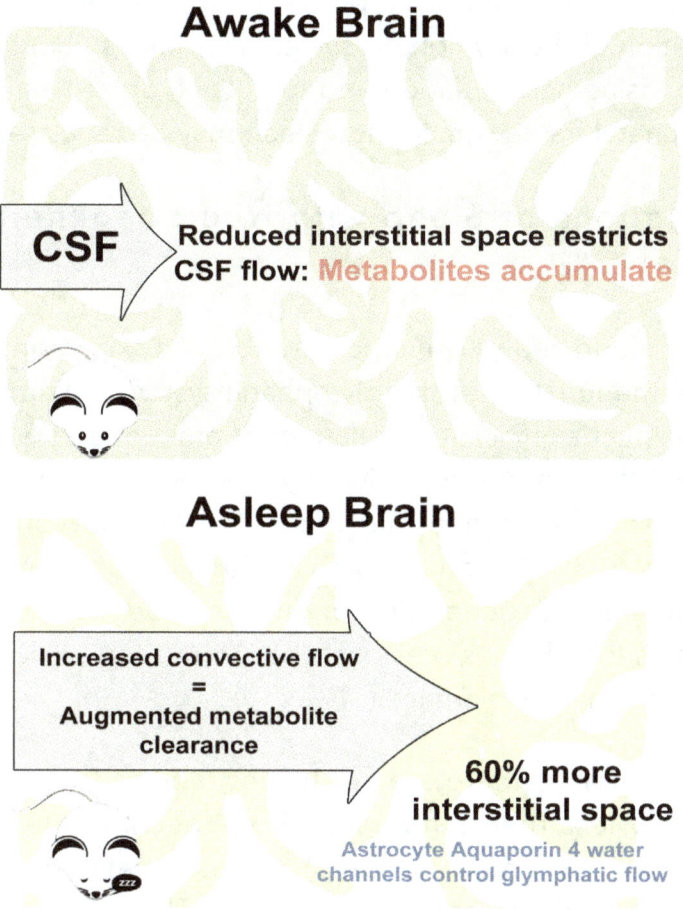

Figure 11.2   Sleep increases glymphatic flow by increasing interstitial space. (Top) In the awake brain, reduced interstitial space restricts CSF flow, allowing metabolites to accumulate. In the sleeping brain, aquaporin 4-mediated changes in astrocyte cell volume, increases interstitial space to increase CSF access. (Bottom) Resulting increased convective flow augments metabolite clearance.

existing noninvasive interventions. In another report, the same group suggests lumbar intrathecal (spinal canal) injection of tracers visible by contrast-enhanced MRI for studies in humans (14) (6).

There is potential benefit in retrospectively reviewing earlier results in light of this new report. For example, accumulation of adenosine in the brain was reported to drive induction of sleep (15). Modulation of the glymphatic system provides a natural mechanism for adenosine accumulation and removal (16), as well a potentially more general framework for how sleep may be regulated.

Correlations of Sleep with Neurodegeneration and Aging

If sleep really drives clearance of toxic biomolecules by altering the function of the glymphatic system in humans, then we would expect significant correlations of amount and quality of sleep with human neurodegenerative disease and longevity. And indeed there is some evidence that this is the case beyond the strong associations seen with lack of sleep and mortality/neurodegeneration in flies and rodents (11) (12). Most relevant is that IFS has previously been reported to decline from 19 to 13% in young vs old mice (17).

In humans there are consistent reports that both too little (generally less than 7 hours of sleep) and too much (generally greater than 9 hours of sleep) are associated with increased mortality and decreased quality of life. This correlation occurs in older adults, but earlier sleep disruptions may play a role in longevity. Cause and effect are notoriously difficult to establish in such studies, and differences in endpoints and methodology make such studies problematic for effective meta-analyses. Many of the studies were focused on connections to cardiovascular disease, which has been correlated with altered sleep (18). Likely, death occurs before typical onset of neurodegenerative diseases. For direct support of sleep-associated glymphatic clearance, more studies focused on the role of sleep on neurodegeneration and cognition are needed (19), although those that have been performed are consistent with sleep playing an important role (20).

The evidence suggests that organ systems other than the brain are more closely tied to sleep dysfunction and mortality, however, at a minimum, cognitive ability deteriorates with altered sleep. Regardless, the association of mortality with too much sleep is especially interesting, suggesting a "Goldilocks" effect where just the correct amount of sleep is needed to avoid problems, and the existence of trade-offs in sleep versus waking physiology. We speculate that brain function may suffer from too much time spent "house cleaning", as well as too little.

What about more direct evidence for a role of sleep in preventing neurodegenerative diseases such as Alzheimer's in humans?

A small clinical study (~70 people) showed that subjects with shorter sleep duration or shorter sleep quality have an associated greater Aβ burden, measured by a non-invasive probe for amyloid (Pittsburgh compound B) positron emission tomography (21). This result is completely consistent with the potential role of sleep-associated glymphatic waste removal. Although this is the best example of a potentially direct connection of amyloid plaque in humans with sleep duration and quality, a significant set of data are consistent with this result (see the review "A Hypnotic Hypothesis of Alzheimer's Disease" (20) for more details).

Changes in sleep pattern may predict Alzheimer's disease. In a 9 year study, 214 dementia-free Swedish adults (>= 75 years old, mean age 83.4 years old) were followed. 40% of the subjects reported a change in sleep pattern. Between years 6 and 9, 28.5% of were diagnosed with dementia, 22.0% of whom had AD. Reduced sleep was associated with a 75% increased all-cause dementia risk (hazard ratio: 1.75; 95% confidence interval: 1.04-2.93; p = 0.035) and double the risk of AD (hazard ratio: 2.01; 95%

confidence interval: 1.12-3.61; p = 0.019) after adjusting for age, gender, and education. Only adjusting for depression, a frequent correlate of sleeplessness, caused a loss of significance. Lifestyle, vascular status and apoE status did not alter the results. Sleeping pills proved ineffective (22). In these types of correlative studies, cause and effect is always somewhat unclear, but it is interesting that sleeping pills, which presumably would induce an anesthesia-like state similar to that seen in the mouse studies were ineffective. Because sleeping pills may not result in quality sleep, more careful investigation of sleeping stage (NREM stages, REM, etc) and glymphatic function may prove enlightening.

Even more intriguing is a recent report that sleep is a significant attenuator of risk for Alzheimer's disease for people carrying the Apolipoprotein E (ApoE) ε4 allele. The ApoE ε4 allele increases risk for patients to develop AD by approximately two-fold for people with one copy, and as much as 10 fold for those carrying two copies by age 79. Risk for those carrying two copies of the ApoE ε4 allele may be a high as 70% by age 79. In this study 698 community-dwelling older adults without dementia (mean age, 81.7 years; 77% women) were followed for six years. Subjects were tested for ApoE status, sleep quality/duration by 10 days of actigraphy (which uses a wrist watch like device to measure movement and activity). During the follow-up period, 98 individuals developed AD. Better sleep quality attenuated the effect of the ApoE ε4 allele on the risk of incident AD (hazard ratio, 0.67; 95% CI, 0.46-0.97; P = .04 per allele per 1-SD increase in sleep quality/duration) and on the rate of cognitive decline. From post-mortem examination of the 201 people who died during the study, better sleep quality attenuated the effect of the ApoE ε4 allele on neurofibrillary tangle density, which accounted for the effect of sleep quality on the association between APOE genotype and cognition before death. These data support the possibility that supporting improved sleep may reduce the risk of AD and development of neurofibrillary tangles in APOE ε4+ individuals (23).

Medical Implications

The medical implications of these results are profound. If the glymphatic system plays as important role as the Xie et al results suggest, then glymphatic dysfunction may play a hitherto unsuspected role in the pathogenesis of neurodegenerative diseases as well as the maintenance of cognition. Such mechanisms would be critical to unravel and likely work is ongoing to achieve this goal.

If improved sleep can help prevent or delay neurodegenerative diseases, especially Alzheimer's, then a public health effort should be made to achieve better sleep, especially among middle-aged and older people. A special effort should be made for people carrying the ApoE e4 alleles, as it appears that good sleep reduces the risk of carriers by 33%. A Japanese study suggested that a short nap (30 minutes between 1300 and 1500 hours) in which the subject does not go into a deep sleep (which is actually counterproductive) and moderate exercise such as walking could positively impact sleep quality (24). In fact short naps may actually help prevent AD (25). A 4 week trial of this intervention demonstrated greatly improved sleep efficiency, as well as physical and mental health (24).

Developing new drugs that stimulate removal of waste products in the brain is an exciting possibility. However, it is possible that some such drugs may already exist. Because adrenergic antagonists recapitulated the effects of sleep in awake animals, perhaps adrenergic inhibitors could prove useful. Perhaps beta adrenergic blockers presently used to control systolic high blood pressure should be evaluated. Intriguingly, a recent study subjects taking beta adrenergic blockers (BB) showed reduced risk of AD (26). However, in this case <u>all treatments</u> that reduced blood pressure decreased the risk of AD including diuretics, angiotensin-1 receptor blockers (ARB), angiotensin-converting enzyme inhibitors (ACE-I) and calcium channel blockers (CCB). Interestingly, the beneficial effect was in addition to or independent of mean systolic blood pressure. However, in an earlier study of patients with incipient AD, beta-blockers, but not ACE inhibitors, were associated with a reduced rate of the progression of AD (27).

Conclusions

The existence of a sleep-driven convective system in the brain to remove metabolic waste products has profound consequences for maintenance of brain function during aging. Further characterization of the glymphatic system in humans may lead to new therapies and methods to prevent neurodegenerative diseases. A public health initiative to ensure enough, proper sleep in middle-aged and older people may prove useful in preventing AD, especially among ApoE ε4 carriers. The benefits of sufficient sleep to cognition and quality of life have been evident since (and before) the times of Hippocrates, but these benefits may be greater than originally imagined!

Acknowledgements

The authors wish to acknowledge the graphic design work of Ms. Jasmine Larrick.

References

[1] Tanzi RE. The Genetics of Alzheimer Disease. Cold Spring Harb Perspect Med 2012;2:a006296.
[2] Yoon S-S, Jo SA. Mechanisms of Amyloid-beta Peptide Clearance: Potential Therapeutic Targets for Alzheimer's Disease. Biomol Ther 2012;20:245–55.
[3] Chesser AS, Pritchard SM, Johnson GVW. Tau Clearance Mechanisms and Their Possible Role in the Pathogenesis of Alzheimer Disease. Front Neurol 2013;4.
[4] Iliff JJ, Nedergaard M. Is there a cerebral lymphatic system? Stroke J Cereb Circ 2013;44:S93–95.
[5] Iliff JJ, Wang M, Liao Y, Plogg BA, Peng W, Gundersen GA, et al. A Paravascular Pathway Facilitates CSF Flow Through the Brain Parenchyma and the Clearance of Interstitial Solutes, Including Amyloid Beta Sci Transl Med 2012;4:147ra111.
[6] Iliff JJ, Lee H, Yu M, Feng T, Logan J, Nedergaard M, et al. Brain-wide pathway for waste clearance captured by contrast-enhanced MRI. J Clin Invest 2013;123:1299–309.
[7] Xie L, Kang H, Xu Q, Chen MJ, Liao Y, Thiyagarajan M, et al. Sleep drives metabolite clearance from the adult brain. Science 2013;342:373–7.
[8] Stickgold R. Neuroscience: a memory boost while you sleep. Nature 2006;444:559–60.
[9] Malow BA. Sleep Deprivation and Epilepsy. Epilepsy Curr 2004;4:193–5.
[10] Boonstra TW, Stins JF, Daffertshofer A, Beek PJ. Effects of sleep deprivation on neural functioning: an integrative review. Cell Mol Life Sci 2007;64:934–46.
[11] Shaw PJ, Tononi G, Greenspan RJ, Robinson DF. Stress response genes protect against lethal effects of sleep deprivation in Drosophila. Nature 2002;417:287–91.
[12] Rechtschaffen A, Gilliland MA, Bergmann BM, Winter JB. Physiological correlates of prolonged sleep deprivation in rats. Science 1983;221:182–4.
[13] Montagna P, Gambetti P, Cortelli P, Lugaresi E. Familial and sporadic fatal insomnia. Lancet Neurol 2003;2:167–76.
[14] Yang L, Kress BT, Weber HJ, Thiyagarajan M, Wang B, Deane R, et al. Evaluating glymphatic pathway function utilizing clinically relevant intrathecal infusion of CSF tracer. J Transl Med 2013;11:107.
[15] Porkka-Heiskanen T, Strecker RE, Thakkar M, Bj?Barkum AA, Greene RW, McCarley RW. Adenosine: A Mediator of the Sleep-Inducing Effects of Prolonged Wakefulness. Science 1997;276:1265–8.
[16] Herculano-Houzel S. Sleep It Out. Science 2013;342:316–7.

[17] Sykova E, Vorisek I, Antonova T, Mazel T, Meyer-Luehmann M, Jucker M, et al. Changes in extracellular space size and geometry in APP23 transgenic mice: A model of Alzheimer's disease. Proc Natl Acad Sci U S A 2005;102:479–84.

[18] Ikehara S, Iso H, Date C, Kikuchi S, Watanabe Y, Wada Y, et al. Association of Sleep Duration with Mortality from Cardiovascular Disease and Other Causes for Japanese Men and Women: the JACC Study. Sleep 2009;32:295–301.

[19] Virta JJ, Heikkilä K, Perola M, Koskenvuo M, Räihä I, Rinne JO, et al. Midlife sleep characteristics associated with late life cognitive function. Sleep 2013;36:1533–41.

[20] Clark CN, Warren JD. A Hypnic Hypothesis of Alzheimer's Disease. Neurodegener Dis 2013;12:165–76.

[21] Spira AP, Gamaldo AA, An Y, Wu MN, Simonsick EM, Bilgel M, et al. Self-reported Sleep and β-Amyloid Deposition in Community-Dwelling Older Adults. JAMA Neurol 2013.

[22] Hahn EA, Wang H-X, Andel R, Fratiglioni L. A Change in Sleep Pattern May Predict Alzheimer Disease. Am J Geriatr Psychiatry 2013.

[23] Lim ASP, Yu L, Kowgier M, Schneider JA, Buchman AS, Bennett DA. Modification of the Relationship of the Apolipoprotein E ε4 Allele to the Risk of Alzheimer Disease and Neurofibrillary Tangle Density by Sleep. JAMA Neurol 2013.

[24] Tanaka H, Shirakawa S. Sleep health, lifestyle and mental health in the Japanese elderly: ensuring sleep to promote a healthy brain and mind. J Psychosom Res 2004;56:465–77.

[25] Asada T, Motonaga T, Yamagata Z, Uno M, Takahashi K. Associations between retrospectively recalled napping behavior and later development of Alzheimer's disease: association with APOE genotypes. Sleep 2000;23:629–34.

[26] Yasar S, Xia J, Yao W, Furberg CD, Xue Q-L, Mercado CI, et al. Antihypertensive drugs decrease risk of Alzheimer disease: Ginkgo Evaluation of Memory Study. Neurology 2013;81:896–903.

[27] Rosenberg PB, Mielke MM, Tschanz J, Cook L, Stat M, Corcoran C, et al. Effects of Cardiovascular Medications on Rate of Functional Decline in Alzheimer Disease. Am J Geriatr Psychiatry Off J Am Assoc Geriatr Psychiatry 2008;16:883–92.

Section IV

Altering epigenetics, differentiation and engineering rejuvenation

Chapter 12

Overcoming the aging systemic milieu to restore neural stem cell function

As mammals age, the rate of neurogenesis in the brain declines with a concomitant reduction in cognitive ability. Recent data suggest that plasma-borne factors are responsible for inhibition of neurogenesis. When the circulatory systems of old and young mice are connected, the old mice experience increased neurogenesis and the young mice exhibit less neurogenesis, suggesting the importance of systemic circulating factors. Chemokine CCL11/eotaxin has been identified as a factor that increases with aging. Injections of CCL11 inhibit neurogenesis in young mice, an effect likely mediated by CCR3 receptors on neural stem cells. Identification of a specific factor that plays a causative role in stem cell dysfunction in aging is consistent with data showing that transforming growth factor-beta (TGF-beta) inhibits satellite cell-mediated repair. Together, these data suggest that the systemic milieu plays a critical role in the aging of adult stem cells. Because adult stem cells help maintain homeostasis by providing the possibility of replacing metabolically damaged differentiated cells, aging of the systemic milieu and stem cell niches may drive functional decline during aging. The identification of a specific systemic change suggests that aging is more amenable to therapeutic modulation than work on global metabolism-derived damage and cellular senescence implies.

Stem Cell Aging Has Been Linked to Altered Stem Cell Niches and the Systemic Milieu

Aging organisms gradually lose the ability to maintain

homeostasis. Differentiated cells, especially those that are post-mitotic, have a limited ability to maintain their state over time and need to be replaced. Multicellular organisms have evolved rejuvenative mechanisms based on populations of progenitor and adult stem cells that act as a reservoir for the production of newly differentiated cells to help maintain organismal integrity. Although these stem cell populations are insufficient to replace all aging differentiated cells completely in most animals, they likely help to extend longevity relative to organisms that have fewer or lack such reservoirs. However, recent work indicates that adult stem cell function declines with organismal age and that this loss is often associated with alteration of the stem cell niche and systemic milieu that helps maintain stem cell function (1). Perhaps this is not terribly surprising given that the stem cell niches and the systemic milieu are themselves composed of differentiated cells subject to aging. But why don't stem cells simply replace all aging differentiated cells to avoid deterioration of their milieu? The answer is that evolution only optimizes successful reproduction and not necessarily organismal integrity. Evolution has not selected for omnipotent stem cell-based repair systems for most animals, with the possible exception of certain asexual organisms such as hydra and some planarian species, which require such repair systems to reproduce. In the absence of such efficient repair systems, a gradual decline in the function of unreplaced or incompletely replaced aging cells eventually compromises the function of near and distant stem cells, as well as other differentiated cells, leading to loss of homeostasis.

Mechanisms of Stem Cell Aging

Loss of stem cell function in aging has been hypothesized to be due to several mechanisms: Depletion due to differentiation, loss of capacity for self-renewal or senescence, loss of lineage specificity (ability to make correct, fully functional progeny), or malignant transformation. These dysfunctional states can result from altered intrinsic metabolism or altered extrinsic influences (e.g., changes in trophic factors, inflammatory factors, etc). Among the best-studied examples of decreased stem cell function are the shift to an adipogenic program in adult myoblasts (2) and in tendon-derived stem/ progenitor cells (3) with age. A good example of stem cell depletion is loss of melanocyte stem cells by genotoxic stress resulting in gray hairs (4). The best-understood mechanism of stem cell dysfunction in which the environment plays a clear role is the decline of skeletal muscle stem cell (satellite cell) function with age. Satellite cells' capability to repair damaged skeletal muscle is controlled by a balance between notch and wnt signaling (5) and is gradually lost with age. However, the loss is reversible. Connecting the circulatory system of a young mouse with an old mouse restores activity of old satellite cells (6). Although the precise factors from plasma that cause this effect are unknown, it has been established that increased transforming growth factor-beta (TGF-beta) in the local environment of the satellite cells can block satellite cell function (7) and it has been hypothesized that a systemic TGF-beta antagonist neutralizes TGF-beta in young but not old animals (8). These results support the paradigm of age-associated stem cell dysfunction by defined changes in the systemic milieu.

Cytokine CCL11/Eotaxin-1 Increases with Age and Inhibits Neurogenesis

In an important study that extends our understanding of stem cell function during aging, Villeda *et al.* show that cytokine CCL11/eotaxin-1 serum levels increase with advancing age, impairing neurogenesis and cognitive function (9). Adult neurogenesis, which occurs in blood vessel-rich neurogenic niches in the subventricular zone (SVZ) and the subgranular zone (SGZ) of the hippocampus, may play a role in learning and cognition in mammals, although there is some controversy in the literature as to exactly which cognitive processes are affected (10–12). First, Villeda *et al.* (9) showed that in their mouse cohort, as expected, old mice had increased neuroinflammation, decreased neurogenesis, and synaptic plasticity and deficits in cognitive behaviors (contextual fear conditioning and spatial learning in a radial arm water maze [RAWM]).

To test the hypothesis that age-related changes in the systemic milieu might cause decreased neurogenesis and cognitive loss, mice were subjected to heterochronic (young–old) parabiosis whereby the circulatory systems of old and young mice are surgically linked. In the young parabionts, there was a decrease of SOX-2 (also called SRY [sex determining region Y]-box 2)-expressing neural precursors, doublecortin (Dcx)-expressing newly born neurons, and overall cell proliferation as assessed by bromodeoxyuridine (BrdU) incorporation compared to young–young parabiotic pairs. Reciprocally, there was an increase in neurogenesis as measured by these parameters in the old parabionts compared to old–old parabiotic–paired controls. Taken together, these results support the hypothesis that the systemic milieu modulates neurogenesis during aging. The authors ruled out direct migration of cells into the brains of the parabionts by using green fluorescent protein (GFP)-expressing transgenic mice as one of the pair of connected animals. Consistent with the idea that blood from old mice makes soluble inhibitory factors, injection of plasma from old mice inhibited neurogenesis in young mice. Furthermore, young mice receiving old plasma showed alterations in two narrow tests of cognition: Decreased freezing in contextual, but not cued, memory tests, and impaired learning and memory for platform location in the RAWM test.

To determine the identity of the soluble factors, Villeda *et al.* used enzyme-linked immunosorbent assay (ELISA) analysis to test the levels of 66 cytokines, chemokines, and other factors in heterochronic parabionts and in unpaired old mice. They determined that CCL2, CCL11, CCL12, CCL19, haptoglobin, and beta-2-microglobulin were elevated in both groups of animals. The authors then focused on CCL11, which was also observed to show increased levels with age in analysis of random human plasma samples. CCL11 was an auspicious choice because it was previously shown to inhibit neurogenesis in human neural precursors in culture (13), although Villeda *et al.* (9) were apparently unaware of this work. Consistent with this prior work, CCL11 inhibited neurogenesis of mouse primary neural stem/progenitor cells (NPCs) as measured by neurosphere formation and neural differentiation of the human nTERA cell line in culture. Injections of CCL11 into young mice inhibited neurogenesis, but could be rescued by co-administration of CCL11 neutralizing antibody. CCL11-injected young mice showed similar cognitive defects to young mice treated with plasma.

The evidence taken together suggests that systemic CCL11 is upregulated during aging and plays a role in inhibiting neurogenesis, resulting in some decline in cognitive function. The authors claim that this is the first report of a specific systemic factor associated with aging that affects adult stem cell function. However, it is important to note that the authors did not report that injection of anti-CCL11 neutralizing antibodies could restore neurogenesis in old animals. It is unclear whether this experiment had been attempted, but it would significantly strengthen the case for CCL11 being the critical serum factor that mediates systemic inhibition of neurogenesis during aging. Because the authors suggest that blood be more extensively characterized for other factors that may contribute to both inhibition and promotion of neurogenesis, it is likely that multiple factors play a role in inhibition.

Effects of Integration Between the Immune System and Central Nervous System

Neurogenesis has been previously shown to be sensitive to inhibition by inflammation. Increased inflammation is a hallmark of aging in mammals, thus decreased neurogenesis has been hypothesized to be caused by aging-associated inflammation. The work of Villeda *et al.* is not inconsistent with this hypothesis, but the involvement of CCL11 suggests that the connection between altered immune system function and neurogenesis may be more subtle. Several recent reports support the idea that the immune system and the brain may be more tightly integrated than previously appreciated. For example, meningeal T cells support learning, perhaps through secretion of interleukin-4 (IL-4), which maintains a state of "muted inflammation" (14,15). CCL11 can inhibit IL-4 activity (16) potentially increasing inflammation in the critical brain regions (15) although as mentioned above CCL11 can directly affect NPCs, so this effect may be at most a secondary one. Another instructive example is that the ratio of CD4 to CD8 T cells in heterogeneous mouse strains directly correlates with the levels of neurogenesis seen in any particular strain, suggesting a deep connection between T cell function and neurogenesis (12). A potentially intimate connection between the immune system and the central nervous system would add to the number of potential ways by which aging-associated decline of function in one cell population could critically impact another cell population in mammals.

Medical Implications

Although neurogenesis plays a role in memory and cognitive function in mice, it is only part of a more complex set of mechanisms to ensure plasticity (11) There are other aspects of brain function that decline during aging (17) that may be addressed by potential intervention (18,19). To the extent that neurogenesis is a potential target to improve age-related decline in cognition, there are several reported means to achieve increased neurogenesis: Supplementation with curcumin (20), and apigenin (21), exercise (22,23), and dietary restriction (24–26). The evidence that these may be helpful in humans comes from animal models and must be considered preliminary.

In situations in which inflammation blocks neurogenesis, such as inflammation associated with cranial radiation therapy or lipopolysaccharide injection, indomethacin has been reported to restore neurogenesis (27) Interestingly, indomethacin has been reported to act similarly to CCL11 on eosinophils at low doses, but decreases expression of CCR3, the CCL11 receptor, at high concentrations (28). Targeting CCL11 function must be considered completely speculative at this time, because it may be necessary to inhibit the function of several plasma-borne factors to effectively induce neurogenesis in older people. Substances that have b9een reported to inhibit CCL11 function include nobiletin (29) and 7,4-dihydroxy flavone from *Glycyrrhiza uralensis* (30,31). However, there is no available evidence that these molecules will induce neurogenesis, although it is possible that they would be able to achieve some beneficial effect, especially in conjunction with treatments that appear to stimulate neurogenesis.

Conclusions

Subtle specific changes in the stem cell milieu, among them altered serum levels of specific factors such as CCL11, may play a more significant functional role in normal human aging than more dysfunctional cell states, such as cellular and replicative senescence, that have been hypothesized to play critical roles in aging through promotion of increased inflammation (32). The existence of potentially important specific age-associated molecular changes in stem cells opens the door to the discovery and development of interventions to slow aging.

References

1. Liu L, Rando TA. Manifestations and mechanisms of stem cell aging. J Cell Biol 2011;193:257–266.
2. Taylor-Jones JM, McGehee RE, Rando TA, Lecka-Czernik B, Lipschitz DA, Peterson CA. Activation of an adipogenic program in adult myoblasts with age. Mech. Ageing Dev 2002;123:649–661.
3. Zhou Z, Akinbiyi T, Xu L, Ramcharan M, Leong DJ, Ros SJ, Colvin AC, Schaffler

MB, Majeska RJ, Flatow EL, Sun HB. Tendon-derived stem/progenitor cell aging: Defective self-renewal and altered fate. Aging Cell 2010;9:911–915.

4. Inomata K, Aoto T, Binh NT, Okamoto N, Tanimura S, Wakayama T, Iseki S, Hara E, Masunaga T, Shimizu H, Nishimura EK. Genotoxic stress abrogates renewal of melanocyte stem cells by triggering their differentiation. Cell 2009;137:1088–1099.

5. Brack AS, Conboy MJ, Roy S, Lee M, Kuo CJ, Keller C, Rando TA. Increased wnt signaling during aging alters muscle stem cell fate and increases fibrosis. Science 2007;317: 807–810.

6. Conboy IM, Conboy MJ, Wagers AJ, Girma ER, Weissman IL, Rando TA. Rejuvenation of aged progenitor cells by exposure to a young systemic environment. Nature 2005;433: 760–764.

7. Carlson ME, Hsu M, Conboy IM. Imbalance between psmad3 and notch induces cdk inhibitors in old muscle stem cells. Nature 2008;454:528–532.

8. Carlson ME, Conboy MJ, Hsu M, Barchas L, Jeong J, Agrawal A, Mikels AJ, Agrawal S, Schaffer DV, Conboy IM. Relative roles of tgf-b1 and wnt in the systemic regulation and aging of satellite cell responses. Aging Cell 2009;8:676–689.

9. Villeda SA, Luo J, Mosher KI, Zou B, Britschgi M, Bieri G, Stan TM, Fainberg N, Ding Z, Eggel A, Lucin KM, Czirr E, Park J-S, Couillard-Despres S, Aigner L, Li G, Peskind ER, Kaye JA, Quinn JF, Galasko DR, Xie XS, Rando TA, Wyss-Coray T. The ageing systemic milieu negatively regulates neurogenesis and cognitive function. Nature 2011;477: 90–94.

10. Leuner B, Gould E, Shors TJ. Is there a link between adult neurogenesis and learning? Hippocampus 2006;16:216–224. 11. Deng W, Aimone JB, Gage FH, others. New neurons and new memories: How does adult hippocampal neurogenesis affect learning and memory. Nat Rev Neurosci 2010;11:339– 350.

12. Huang G-J, Smith AL, Gray DHD, Cosgrove C, Singer BH, Edwards A, Sim S, Parent JM, Johnsen A, Mott R, Mathis D, Klenerman P, Benoist C, Flint J. A genetic and functional relationship between t cells and cellular proliferation in the adult hippocampus. PLoS Biol 2010;8:e1000561.

13. Krathwohl MD, Kaiser JL. Chemokines promote quiescence and survival of human neural progenitor cells. Stem Cells 2004;22:109–118.

14. Derecki NC, Cardani AN, Yang CH, Quinnies KM, Crihfield A, Lynch KR, Kipnis J. Regulation of learning and memory by meningeal immunity: A key role for il-4. J. Exp. Med. 2010;207:1067–1080.

15. Ransohoff RM. Ageing: Blood ties. Nature 2011;477:41–42. 16. Stevenson NJ, Addley MR, Ryan EJ, Boyd CR, Carroll HP, Paunovic V, Bursill CA, Miller HC, Channon KM, McClurg AE, Armstrong MA, Coulter WA, Greaves DR, Johnston JA. Ccl11 blocks il-4 and gm-csf signaling in hematopoietic cells and hinders dendritic cell differentiation via suppressor of cytokine signaling expression. J Leukoc Biol 2009 ;85: 289–297.

17. Toescu EC. Normal brain ageing: Models and mechanisms. Philos Trans R Soc Lond. B Biol Sci 2005; 360:2347–2354.
18. Kirkwood TBL. Global aging and the brain. Nutr Rev 2010;68(Suppl 2):S65–S9.
19. Mendelsohn AR, Larrick JW. Reversing age-related decline in working memory. Rejuvenation Res 2011;14:557–559.
20. Kim SJ, Son TG, Park HR, Park M, Kim M-S, Kim HS, Chung HY, Mattson MP, Lee J. Curcumin stimulates proliferation of embryonic neural progenitor cells and neurogenesis in the adult hippocampus. J Biol Chem 2008;283: 14497–14505.
21. Taupin P. Apigenin and related compounds stimulate adult neurogenesis. Exp Opin Therapeut Patents 2009;19:523–527.
22. van Praag H. Exercise and the brain: Something to chew on. Trends Neurosci 2009;32:283–290.
23. van Praag H, Kempermann G, Gage FH. Running increases cell proliferation and neurogenesis in the adult mouse dentate gyrus. Nat Neurosci 1999;2:266–270.
24. Lee J, Duan W, Long JM, Ingram DK, Mattson MP. Dietary restriction increases the number of newly generated neural cells, and induces bdnf expression, in the dentate gyrus of rats. J Mol Neurosci 2000;15:99–108.
25. Lee J, Seroogy KB, Mattson MP. Dietary restriction enhances neurotrophin expression and neurogenesis in the hippocampus of adult mice. J Neurochem 2002;80:539–547.
26. Lee J, Duan W, Mattson M. Evidence that brain-derived neurotrophic factor is required for basal neurogenesis and mediates, in part, the enhancement of neurogenesis by dietary restriction in the hippocampus of adult mice. J Neurochem 2002;82:1367–1375.
27. Monje ML, Toda H, Palmer TD. Inflammatory blockade restores adult hippocampal neurogenesis. Science 2003;302: 1760–1765.
28. Stubbs VEL, Schratl P, Hartnell A, Williams TJ, Peskar BA, Heinemann A, Sabroe I. Indomethacin causes prostaglandin d2-like and eotaxin-like selective responses in eosinophils and basophils. J Biol Chem 2002;277:26012–26020.
29. Wu Y-Q, Zhou C-H, Tao J, Li S-N. Antagonistic effects of nobiletin, a polymethoxyflavonoid, on eosinophilic airway inflammation of asthmatic rats and relevant mechanisms. Life Sci 2006;78:2689–2696.
30. Bolleddula J, Doddaga S, Alexandra C, Goldfarb J, Wang R, Li X. Eotaxin-1 inhibition by 7,4-dihydroxy flavone isolated from glycyrrhiza uralensis. J Allergy Clin Immunol 2008;121: S267–S267.
31. Jayaprakasam B, Doddaga S, Wang R, Holmes D, Goldfarb J, Li X-M. Licorice flavonoids inhibit eotaxin-1 secretion by human fetal lung fibroblast in vitro. J Agric Food Chem 2009;57:820–825.
32. Freund A, Orjalo AV, Desprez P-Y, Campisi J. Inflammatory networks during cellular senescence: Causes and consequences. Trends Mol Med 2010;16:238–246.

Chapter 13

Epigenetic-mediated decline in synaptic plasticity during aging

Cognitive decline observed in aging mammals is associated with decreased long-term synaptic plasticity, especially long-term potentiation (LTP). Recent work has uncovered a connection between LTP, histone acetylation, and brain-derived neurotrophic factor (BDNF)/neurotrophin receptor B (trkB) signaling. LTP, histone acetylation, and BDNF/trkB signaling decrease in old animals, Because an apparent positive feedback loop links these processes, treatment with histone deacetylase inhibitors or a trkB agonist restores LTP in the hippocampus of old animals. These results coupled with exciting work on histone methylation and life span in Caneorhabditis elegans suggest that epigenetic changes may play a significant role in aging. Such dysfunctional epigenetic pathways may provide novel targets for cognitive enhancing therapeutics.

Introduction

Cognitive decline with advancing age has been observed in animals and humans. Long-term potentiation (LTP), long-lasting activity between neurons following electrical stimulation, is thought to underlie learning and memory. Decreased long-term synaptic plasticity, especially LTP, and loss of dendritic spines in the hippocampus are among the specific neurological changes thought to cause reduced cognitive ability. Chromatin remodeling through histone acetylation at lysine residues is a fundamental cellular regulatory associated with the activation of transcription, enhances learning and memory in rodent models (1–4). Histone acetylation is controlled enzymatically by the relative activities of histone acetyltransferases (HAT) and deacetylases (HDAC) (**Figure 13.1**).

Although altered chromatin structure, including altered histone acetylation, has been reported to play an important role in replicative senescence and chronological aging in yeast, and histone H3 lysine-4 trimethylation can affect life span in the nematode *Caaenorhabditis elegans* (5), the role of chromatin remodeling in mammalian aging has been less clearly defined (6). Altered histone acetylation patterns have been observed in Hutchinson–Gilford progeria syndrome, which results from a mutation in lamin A (6). However, a critical role for histone acetylation in mammalian aging has not been established. Nevertheless, because histone acetylation is dynamic and intimately involved in processes susceptible to aging-associated dysfunction, such as the maintenance of stable cell differentiation, memory, and learning, there are good reasons to suspect that histone acetylation plays an important role in mammalian aging. In an important paper, Zeng *et al.* (7) provide evidence for these suspicions by demonstrating that reduced histone acetylation causes deficits in LTP with loss of dendritic spikes in the hippocampus of aged Fisher 344 rats.

Loss of Synaptic Plasticity During AgingIs Due to Altered Histone Acetylation and Altered Brain-Derived Neurotrophic Factor/Neurotrophin Receptor B Signaling

Zeng *et al.* extended previous work that showed that reduced histone acetylation with decreased levels of neurotrophic factor brain-derived neurotrophic factor (BDNF) or its receptor neurotrophin receptor B (trkB) can result in decreased LTP, memory, and learning by observing decreased levels of histone acetylation, BDNF, and trkB in old rats. They reported that agents that stimulate histone acetylation or trkB signaling restore signaling are closely linked (**Figure 13.1**), such that restoring normal levels of either one, restores normal levels of the other. Experiments to quantify LTP were performed on hippocampal brain slices from young, middle-aged, or old rats subjected to high-frequency (100 Hz) electrical stimulation for 1 sec. Electrical activity for 75 min following stimulation was measured. LTP decreases moderately in middle-aged rats and more strongly in aged rats. Pretreatment with HDAC inhibitors trichostatin A (TSA) or sodium butyrate (SB) for 3 hr restores LTP levels to values near those seen in young animal controls. Reduced numbers of dendritic spines, which serve as postsynaptic structures in 90% of excitatory synapses and contribute to synaptic plasticity, were also restored by TSA or SB treatment. Interestingly, this restorative effect requires transcription and BDNF signaling, because it can be blocked by both the transcription inhibitor actinomycin D and a BDNF sequestering protein. Consistent with these results, histone H3 acetylation at lysine 9 (H3K9) and histone H4 acetylation at lysine 12 (H4K12) were reduced in aged animals and restored by TSA treatment. Decreased histone acetylation was associated with increased levels of histone deacetylase HDAC2, but not HDAC1 and decreased cyclic-AMP response element binding protein (CREB) binding protein (CBP), a transcriptional coactivator with HAT activity. HDAC2 is known to negatively regulate memory formation (8,9).

Because BDNF was necessary for LTP restoration, Zeng examined pro-BDNF and trkB protein expression and found age-associated decreases that could be restored by TSA or SB treatment. Chromatin immunoprecipitation (ChIP) was used to examine the specific histone acetylation of the BDNF promoter regions. TSA treatment of old rats was seen to increase H3K9 in three out of four promoter regions and increase H4K12 in one region with reduced histone acetylation and downregulation of BDNF during aging. BDNF/trkB regulated signaling cascades Ca++/calmodulin-dependent protein kinase II (CaMKII) and extracellular-signal-related kinase (Erk) (**Figure 13.1**) play important roles in LTP and learning (10,11). In old rats, total CaMKII and Erk expression levels were unaffected, as determined by western blots, but the levels of highly active, phosphorylated CaMKII and Erk were strongly reduced. Phosphorylated CaMKII and Erk levels are restored with

TSA treatment. BDNF is necessary for TSA-mediated increases in phosphorylated CaMKII and Erk as well as for increased histone acetylation and trkB/BDNF expression.

CaMKII plays a significant role in early-phase LTP and promotes dendritic spine growth and synapse formation. Erk promotes late-phase LTP, persistent synaptic plasticityn (12,13) and phosphorylation of CBP, increasing its intrinsic HAT activity (14). The authors hypothesize the existence of a positive feedback loop whereby HDAC inhibition increases BDNF expression, which in turn increases histone acetylation and BDNF/trkB signaling (**Figure 13.1**). Zeng *et al.* treated hippocampal slices from old rats with 7,8-dihydroxyflavone (7,8-DHF), a strong, selective, small molecule agonist of trkB. 7,8-DHF rescued LTP activity in old animals, but did not affect LTP in young animals. Similarly to TSA, 7,8-DHF increased histone acetylation, BDNF, phosphorylated CaMKII and Erk levels, and dendritic spine number, confirming a bidirectional control loop between BDNF/trkB and histone acetylation.7 Interestingly, the authors did not examine the direct effects of BDNF, which has had disappointing results in clinical trials (15), possibly due to pharmacokinetic limitations. However, they have studied the trkB agonist 7,8-DHF that may itself have clinical potential (16). Although Zeng *et al.* only performed their experiments in brain slices and did not directly test learning or memory in animals, there is a strong possibility that they will obtain corroborative results, because prior work supports the use of BDNF (17) and HDAC inhibitors (18) as memory enhancers.

The underlying cause of altered BDNF/trkB expression and activity and histone acetylation is unknown, as are the precise roles neurons and glial cells play in the reported effects (7).

Do alterations in histone acetylation, which can be quite global, affect other regulatory or functional pathways in the hippocampus? Probably. Perhaps histone acetylation is also involved in alpha-2a adrenergic receptor activity, which is also associated with short-term memory loss (19,20). Do the changes result from deterioration of epigenetic control over time due to imperfect evolutionary design, accumulated stress-related changes, or molecular damage? It will be interesting to see whether short-term treatment with 7,8- DHF or HDAC inhibitors restores memory for longer periods of time than a few hours, which might imply that stochastic changes are important. However, the establishment of new homeostasis set points during aging may involve multiple physiological changes that are difficult to reset permanently. Similarly, molecular damage would require cleaning, repair, and/or replacement of damaged macromolecules and cells. In light of these possibilities, it is remarkable that normal physiology and epigenetic regulation can apparently be restored with HDAC inhibitors or 7,8- DHF.

Medical Implications

The use of global HDAC inhibitors to promote memory in the elderly is a possibility. For example, the HDAC class 1 inhibitor valproic acid is Food and Drug Aministration (FDA)-approved as an anticonvulsant and modulates the neurotransmitter gamma-aminobutyric acid (GABA). TSA itself has been studied as an anticancer drug, although there is limited clinical experience. Interestingly, HDAC inhibitors have been reported to have activity against neurodegenerative diseases, including Alzheimer's disease, Huntington's disease, and amyotrophic lateral sclerosis in preclinical animal models (21,22). However, concerns related to the capacity of this class of compounds to modify histone acetylation globally may limit enthusiasm for their development for treatment of impaired cognition in the elderly or for prophylaxis of cognitive decline.

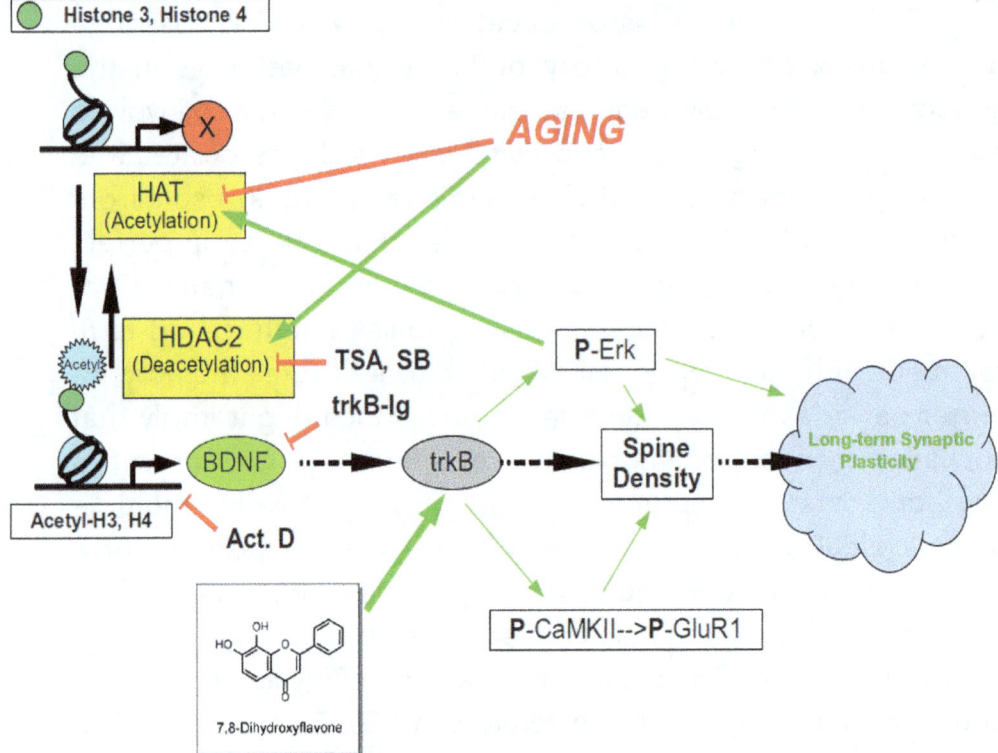

Figure 13.1 Epigenetic enhancement of brain-derived neurotrophic factor (BDNF) signaling modulates synaptic plasticity in aging. Histone/cyclic-AMP response element binding protein (CREB) binding protein (CBP) acetylation by histone acetyltransferases (HAT) facilitates transcription of BDNF. BDNF binds to neurotrophin receptor B (trkB), which activates several second-messenger pathways (e.g., extracellular-signal-related kinase (Erk), Ca2+/calmodulin-dependent protein kinase II [CaMKII], phosphoinositide 3-kinase [PI3K], phospholipase C) that facilitate α-amino-3-hydroxy-5-methyl-4-isoxazolepropionic acid (AMPA) receptor function, resulting in increased dendritic spine density to increase synaptic plasticity. Erk provides a positive feedback loop by enhancing CBP/HAT activity. Compromised synaptic plasticity observed in aging is associated with inhibition of CBP/HAT and stimulation of histone deacetylase (HDAC) with attenuated BDNF synthesis and resultant reduced trkB activation. 7,8-Dihydroxyflavone mimics the effects of BDNF on trkB to enhance hippocampal synaptic function and formation. HDAC inhibitors trichostatin A (TSA) or sodium butyrate (SB) increased transcription of BDNF. Green arrows indicate beneficial pathways. Red arrows indicate inhibitory/deleterious pathways. Act. D, Actinomycin D (generic transcription inhibitor); trkB-Ig, decoy inhibitor of BDNF. [See ref. 7 for experimental details].

The BDNF agonist 7,8-DHF is a promising compound for treatment of defective cognition as well as for seizure, stroke, and Parkinson's disease (16). 7,8-DHF is found in low doses in fruits and vegetables, although levels are unlikely to be in the therapeutic range. Although the pharmacokinetics and toxicology have not been studied, like other flavonoids they are expected to be quite safe. We predict that this compound will be available as a nutraceutical should favorable preclinical results continue to be reported. Deoxygedunin is another naturally derived BDNF agonist with potent neurotrophic activity that may have similar beneficial effects on cognition (23). It should be noted that exercise has been reported to stimulate BDNF levels,24–26 as well as confer other beneficial effects on cognition,27 and potentially provides a simple way to modify the BDNF/trkB/histone acetylation pathways (**Figure 13.1**).

Conclusion

The discovery that aging-related changes in the bidirectionally-linked BDNF/trkB signaling/histone acetylation epigenetic pathways affect LTP extends our knowledge of cellular regulation, neuroscience, and aging. The use of naturally derived trkB receptor agonists, such as 7,8-DHF, represents a promising approach to increase cognitive function and memory in the elderly, although additional preclinical and clinical studies are needed to confirm their effectiveness. This work reinforces an important principle—subtle age-related changes may be reversible after identification of and targeting the relevant molecular defect.

References

1. Vecsey CG, Hawk JD, Lattal KM, Stein JM, Fabian SA, Attner MA, Cabrera SM, McDonough CB, Brindle PK, Abel T, Wood MA. Histone deacetylase inhibitors enhance memory and synaptic plasticity via creb: CBP-dependent transcriptional

activation. J Neurosci 2007;27:6128–6140.

2. Levenson JM, O'Riordan KJ, Brown KD, Trinh MA, Molfese DL, Sweatt JD. Regulation of histone acetylation during memory formation in the hippocampus. J Biol Chem 2004; 279:40545–40559.

3. Fischer A, Sananbenesi F, Wang X, Dobbin M, Tsai L-H. Recovery of learning and memory is associated with chromatin remodeling. Nature 2007;447:178–182.

4. Peleg S, Sananbenesi F, Zovoilis A, Burkhardt S, Bahari-Javan S, Agis-Balboa RC, Cota P, Wittnam JL, Gogol-Doering A, Opitz L, Salinas-Riester G, Dettenhofer M, Kang H, Farinelli L, Chen W, Fischer A. Altered histone acetylation is associated with age-dependent memory impairment in mice. Science 2010;328:753–756.

5. Greer EL, Maures TJ, Ucar D, Hauswirth AG, Mancini E, Lim JP, Benayoun BA, Shi Y, Brunet A. Transgenerational epigenetic inheritance of longevity in Caenorhabditis elegans. Nature 2011;479:365–371.

6. Feser J, Tyler J. Chromatin structure as a mediator of aging. FEBS Lett 2011;585:2041–2048.

7. Zeng Y, Tan M, Kohyama J, Sneddon M, Watson JB, Sun YE, Xie C-W. Epigenetic enhancement of BDNF signaling rescues synaptic plasticity in aging. J Neurosci 2011;31:17800–17810.

8. Salisbury CM, Cravatt BF. Activity-based probes for proteomic profiling of histone deacetylase complexes. Proc Natl Acad Sci USA 2007;104:1171–1176.

9. Guan J-S, Haggarty SJ, Giacometti E, Dannenberg J-H, Joseph N, Gao J, Nieland TJF, Zhou Y, Wang X, Mazitschek R, Bradner JE, DePinho RA, Jaenisch R, Tsai L-H. HDAC2 negatively regulates memory formation and synaptic plasticity. Nature 2009;459:55–60.

10. Minichiello L. TrkB signaling pathways in LTP and learning. Nat Rev Neurosci 2009;10:850–860.

11. Zeng Y, Zhao D, Xie C-W. Neurotrophins enhance CAMKII activity and rescue amyloid-b-induced deficits in hippocampal synaptic plasticity. J Alzheimers Dis 2010;21: 823–831.

12. Impey S, Obrietan K, Wong ST, Poser S, Yano S, Wayman G, Deloulme JC, Chan G, Storm DR. Cross talk between ERK and PKA is required for Ca2+ stimulation of CREB-dependent transcription and erk nuclear translocation. Neuron 1998;21: 869–883.

13. Kelleher III RJ, Govindarajan A, Jung H-Y, Kang H, Tonegawa S. Translational control by MAPK signaling in long-term synaptic plasticity and memory. Cell 2004; 116:467–479.

14. Ait-Si-Ali S, Carlisi D, Ramirez S, Upegui-Gonzalez L-C, Duquet A, Robin P, Rudkin B, Harel-Bellan A, Trouche D. Phosphorylation by p44 MAP kinase/Erk1 stimulates CBP histone acetyl transferase activity in vitro. Biochem Biophys Res Commun 1999;262:157–162.

15. Ochs G, Penn RD, York M, Guess R, Beck M, Tenn J, High J, Malta E, Trauma

M, Sender M, Tokay KV. A phase I/II trial of recombinant methionyl human brain-derived neurotrophic factor administered by intrathecal infusion to patients with amyotrophic lateral sclerosis. Amyotroph Lateral Scler Other Motor Neuron Disord 2000;1: 201–206.

16. Jang S-W, Liu X, Yepes M, Shepherd KR, Miller GW, Liu Y, Wilson WD, Xiao G, Blanchi B, Sun YE, Ye K. A selective trkB agonist with potent neurotrophic activities by 7,8dihydroxyflavone. Proc Natl Acad Sci USA 2010;107: 2687–2692.

17. Cowansage KK, LeDoux JE, Monfils M-H. Brain-derived neurotrophic factor: A dynamic gatekeeper of neural plasticity. Curr Mol Pharmacol 2010;3:12–29.

18. Haggarty SJ, Tsai L-H. Probing the role of HDACs mechanisms of chromatin-mediated neuroplasticity. Neurobiol Learn Mem 2011;96:41–52.

19. Wang M, Gamo NJ, Yang Y, Jin LE, Wang X-J, Laubach M, Mazer JA, Lee D, Arnsten AFT. Neuronal basis of age-related working memory decline. Nature 2011;476:210–213.

20. Mendelsohn AR, Larrick JW. Reversing age-related decline in working memory. Rejuvenation Res 2011;14:557–559.

21. Hahnen E, Hauke J, Trankle C, Eyupoglu IY, Wirth B, Blumcke I. Histone deacetylase inhibitors: Possible implications for neurodegenerative disorders. Expert Opin Investig Drugs 2008;17:169–184.

22. Xu K, Dai X-L, Huang H-C, Jiang Z-F. Targeting HDACs: A promising therapy for Alzheimer's disease. Oxid Med Cell Longev 2011;2011; doi: 10(1)155/2011/143269. Epub 2011 Sep 20.

23. Jang S-W, Liu X, Chan CB, France SA, Sayeed I, Tang W, Lin X, Xiao G, Andero R, Chang Q, Ressler KJ, Ye K. Deoxygedunin, a natural product with potent neurotrophic activity in mice. PLoS One 2010;5.

24. Aguiar AS Jr, Castro AA, Moreira EL, Glaser V, Santos ARS, Tasca CI, Latini A, Prediger RDS. Short bouts of mild intensity physical exercise improve spatial learning and memory in aging rats: Involvement of hippocampal plasticity via Akt, Creb and BDNF signaling. Mech. Aging Dev 2011;132:560–567.

25. Aguiar AS Jr, Speck AE, Prediger RDS, Kapczinski F, Pinho RA. Downhill training up regulates mice hippocampal and striatal brain-derived neurotrophic factor levels. J Neural Transm 2008;115:1251–1255.

26. Heyman E, Gamelin F-X, Goekint M, Piscitelli F, Roelands B, Leclair E, Di Marzo V, Meeusen R. Intense exercise increases circulating endocannabinoid and BDNG levels in humans-possible implications for reward and depression. Psychoneuroendocrinology 2011; Oct 24. [Epub ahead of print].

27. Langdon KD, Corbett D. Improved working memory following novel combinations of physical and cognitive activity. Neurorehabil Neural Repair 2011; 2011 Dec 9. [Epub ahead of print].

Chapter 14

The DNA methylome as a biomarker for epigenetic instability and human aging

Methylation of DNA is intimately involved in control of mammalian/vertebrate gene expression as part of a complex epigenetic regulatory system. We hypothesize that DNA methylation at cytosine–phosphate–guanine sites (CpGs), the "DNA methylome," evolved to increase stability of the differentiated state in somatic vertebrate cells, especially post-mitotic cells, which may have helped to increase longevity. Therefore, the DNA methylome may play a key role in human aging and be an ideal source of biomarkers aging. A new model that links the methylome to chronological age has been reported by Hannum *et al* (1) that accurately predicts age and rate of aging from the DNA methylation state of 71 markers in human blood samples. This model may make possible the development of new anti-aging therapeutics as well as more accurately assess the impact of anti-aging regimens, such as caloric restriction and drugs such as rapamycin. Furthermore, the model reveals information loss with increased age consistent with noise/unstable differentiation-based models of aging. The model may eventually lead to experiments to differentiate the contributions of biomolecular damage and noise/incomplete structural replication during aging.

Introduction

Although the search for well-correlated bio-markers of human aging has been filled with false leads and pitfalls, several markers exhibit some useful correlation with biological aging. For example, in non- metabolizing tissues such as lens, racemization of l- to d-aspartate, as reflected by the d/d + l aspartate (Asp) ratio, increases linearly with age, but is limited to a few tissues with no protein turnover (2,3). A unique metabolic signature has been identified in mice that reliably predicts the age of wild-type mice (4), but it has not yet been shown to apply in humans. Telomere length, which shortens with each cell division in somatic mammalian cells, has been used somewhat successfully as a marker of aging (5). Unfortunately, intrinsic heterogeneity, decay due to environmental stress, and restriction to proliferating cell types make telomere length a less-than-ideal marker for human aging (6). Relative numbers of senescent cells in human skin as determined by the expression of the cyclin-dependent kinase inhibitor p16INK4a has been correlated with human age (7) but the study is very preliminary. Changes in expression of genes involved in DNA repair and metabolism have also been predictive of age in a wide range of organisms and tissues (8–10), although it remains unclear how well statistical models and results derived from these reports would predict human age and aging rate. Nevertheless, correlation of gene expression is a promising approach to the discovery of markers of aging and predictive models. Control of mammalian gene expression involves many levels of epigenetic control, suggesting that epigenetic changes may be useful for deriving biomarkers of aging.

Although epigenetics usually refers to the pattern of biomolecules that contribute to chromatin structure, such as DNA methylation, histone modification, and specific DNA- binding proteins, we can more broadly define epigenetics to include everything that contributes to maintaining a cell differentiation state not directly related to the underlying DNA sequence. In vertebrates, gene expression can be inhibited by high levels of DNA methylation at cytosines (C) present in C–phosphate–guanine (G) (CpG) islands. DNA methylation at CpG islands probably plays a different role in many invertebrates such as *Drosophila* and *Caenorhabditis elegans* where it is almost completely absent. We believe that DNA methylation at CpG islands in vertebrates probably evolved to help stabilize cell differentiation state in long-lived animals, especially those that maintain critical populations of post-mitotic cells. The methylome refers to the complete pattern of DNA methylation in a genome, and may be intimately tied to longevity in vertebrates.

In support of the idea that the methylome is important in aging, an increasing number of studies link changes in the epigenome, especially the methylome, with chronological age (11–17). The most suggestive studies report "epigenetic drift," in which DNA methylation patterns in monozygotic twins diverge with increasing age (14,18,19), which is suggestive of random changes in methylation occurring with time. The mechanisms that underlie epigenetic drift are still unclear. It has been hypothesized that epigenetic drift may result from environmental exposure that promotes stress-induced precise changes in the epigenome (20,21). Alternately, spontaneous chemical alteration of methyl groups, mutation of methylated C to thymine (T) or errors in copying methylation states during DNA replication may lead to epigenetic drift. However, the recent observations that methylation state can change in post-mitotic cells (22) suggests to us that the methylome is in constant flux due to dynamic changes in the epigenome and that epigenetic noise in homeostatic circuits may play a role in causing the random changes in the methylome observed with increasing age. In any case, regardless of precise mechanisms, evidence exists for possible random disruption of the methylome with age. This suggests that an in-depth analysis of the methylome may be a viable source of aging biomarkers, and such biomarkers may be closely linked to fundamental processes that underlie aging.

DNA Methylome in Aging

In an important paper, Hannum et al. (1) derive a quantitative model of aging from DNA CpG methylation patterns using samples from the blood of 656 individuals aged 19–101 years. They calculate a rate of methylome aging by correlating changes in an individual's methylome with chronological age. DNA was analyzed with the Illumina Infinium Human Methylation 450 BeadChip assay (23) which covers 99% of the known open reading frames (ORFs). DNA methylation is detected by a variation of the standard bisulfite assay, which chemically converts unmethylated C's to uracils (U), buts does not affect methylated C's. The differential presence of U is then probed by beads that only bind to either U or C at each target sequence. About 14% of the 485,577 markers had significant associations between the age of the subject and the fraction methylated. Using a small previously published dataset of 40 individuals by another group (24) about 76% (53,670) of the markers retained a statistical significance of $p < 0.05$.

To create a predictive model of aging based on DNA methylation, Elastic Net (25), a sophisticated multivariate regression method, was used to select 71 DNA methylation markers that were most predictive of aging. On the original data set, the model achieved an accuracy of 96% in predicting age with an error of 3.9 years. The authors suggest that it was not coincidental that nearly all of the selected markers were localized to genes associated with aging-related pathologies, DNA damage, and oxidative stress. For example, one of the 71 markers lies within SST (somatostatin), a gene associated with Alzheimer's disease. The model was validated using an independent cohort of 174 people resulting in 91% accuracy and an error of 4.9 years. The model was also validated on a smaller previously published data set (24).

Because there is reasonable evidence that individuals age at different rates, it is expected that any model that correlates biomarkers of aging to chronological age will exhibit errors due to the divergence between biological and chronological age in some individuals. Indeed, 1 individual in the original cohort was 65 years old, but had an apparent biological age of 45 based on the model; another about 50 years old had a methylome that predicted a biological age of 70. To assess whether this variation might reflect true divergence between biological and chronological age, each individual's "apparent methylomic aging rate" (AMAR) was calculated. AMAR was then analyzed for association with body mass index (BMI) and gender. BMI did not correlate with AMAR, which is consistent with an earlier study (26). However, men's methylomes appear to age 4% faster than women's, which is consistent with greater observed longevity for females.

To determine if specific genetic variants in the population could be used to discover biomarkers for chronological age and AMAR, whole-exome sequencing of 252 individuals in the original cohort was performed. A total of 10,694 genetic variants (single-nucleotide polymorphisms [SNPs]) were then associated with the top age-associated methylation markers to find 303 methylation quantitative trait loci (meQTLs). A small subset was then validated. Out of 8 selected genetic variants (SNPs) corresponding to 14 meQTLs, 7 were validated on 322 individuals who were part of the cohort used to derive the original methylome aging model. Among these 7 was a SNP in MBD4, which encodes a protein that actually binds methylated DNA and is associated with genomic instability (27). Of most interest are variants associated with both age and rate of aging (AMAR). SNPs were found in three genes that affected both age and AMAR: NEK4, a kinase involved in cell cycle and cancer; JAKMIP3, a JAK kinase interacting protein associated with glioblastoma; and GTPBP10, a guanosine triphosphate (GTP)-binding protein 10 that associated with altered methylation in STEAP2, which participates in iron and copper uptake. These three variants are expected to be analyzed further in new studies for their impact on longevity.

To determine if the model based on association between age and DNA methylation in blood would hold for other tissues, DNA methylation data from the control category of the Cancer Genome Atlas (28), which included breast, kidney, lung, and skin samples, was analyzed. A most interesting result was obtained. The model directly predicted chronological age, but it needed to be corrected by using a simple linear model to adjust the slope and intercept for the other tissues. In other words, changes in the same tissue-correlated methylation markers did occur in the other tissues, but these changes occurred at a different proportional rate. The authors suggest that this might indicate the possibility of a common molecular clock. However, another interesting possibility is that the epigenome, or perhaps subsets of the epigenome, age at different rates in different tissues. If the latter possibility is true, then models derived from different tissue types that show a direct correlation with age would be expected to have different components. When Hannum *et al.* created de novo age models from the breast, kidney, and lung data, most of the associated markers differed, which is consistent with a differential tissue-aging hypothesis. The only methylation marker shared among all of the models is ELOVL2, fatty acid elongase 2, which was also found to associate with human age in a related DNA methylation-based study (29).

Because DNA methylation near promoters is known to affect transcription, a model was derived linking transcription patterns to age using publicly available data from blood samples from 488 individuals (30) in which the expression of 326 genes associates with age. These genes were more likely to have age-associated methylation markers nearby. A model of aging was built associating methylome age with gene expression. This model, although less accurate than the original methylation model, allowed age prediction based on gene expression and also showed increased rates of aging for men, as would be expected for an accurate model. Further development of a transcription-based model for aging is clearly warranted.

Two calculations were made to confirm whether epigenetic drift occurs. Using a measure based on computing the deviance of each methylation marker as a squared distance from the expected population mean, a total of 27,800 markers showed significant association with age, and of these 99.8% showed increased rather than reduced deviance. Furthermore, for any 1 person, high or low methylome deviance strongly predicted their rate of aging. Using Shannon entropy (31) calculations, which represent loss of information over time as the methylation fraction approaches 50%, similar results were obtained. More than 70% of the methylation markers tended toward 50%, causing a significant loss of methylome information over time. High levels of methylome entropy were correlated with accelerated aging for individuals, and examination of tumor tissue showed 40% more methylome "aging" than matched normal tissue. Although not explicitly calculated by the authors, it is clear that cancer cells show a greater loss of initial epigenetic information than normal cells do, as might be expected for rapidly evolving entities. In our opinion, these data are most consistent with increasing instability of the epigenome and the differentiated state with age that may lie at the heart of aging in multicellular organisms. The relative contributions of damage versus intrinsic loss of information due to incomplete maintenance of genetic/biochemical circuits remains to be elucidated.

Dissection of methylation patterns at the individual cell level would clarify how uniform the observed changes are within any specific tissue type. The creation of a similar model for rodents and other mammals and comparison with humans should deepen our understanding of what determines the rate of aging in mammals. Furthermore, it would be of great interest to see this model extended to tissues and cell types that are quiescent and thought to undergo less variation in methylation. New data find that the methylome may not be as static as originally thought in non-proliferating cells (22), which suggests that age-related changes may also be observed in quiescent cells. Comparing the methylomes of non-proliferating and proliferating cells with respect to changes that correlate with age may help dissect critical differences in the stability of methylation maintenance.

Medical Implications

Interventions to increase life span and health span require accurate assessment of biological age. Such an assessment has not really been possible, perhaps until now. Although validation of the Hannum *et al.*'s methylome model of aging on even larger number of samples and tissue types is necessary to establish clinical relevance, the initial validation of this blood sample–based model on data from different cohorts and tissues from other independent investigations suggests that, even in its present form, the model is likely to be useful.

Assessment of biological age in long-term longitudinal studies such as the Framingham project and the Women's Health Initiative will be helpful in determining how age contributes to pathologies as well as in determining how various biometric factors such as alcohol consumption and smoking may contribute to biological age. The model offers individuals and researchers the opportunity to calculate biological age and rate of aging (at least of an individual's blood cell methylome) before and after potentially beneficial interventions, such as dietary restriction, and treatment with drugs, such as rapamycin and metformin, and supplements, such as curcumin and resveratrol. We expect that tests that measure methylation of the key 71 methylation markers will become available sooner rather than later, even before their clinical efficacy is formally proven.

If the stability of the methylome, and by implication, the epigenome and cell differentiation state, itself is age dependent, then aging may partially derive from profound and subtle information loss in somatic tissue with time. Such information loss may be due to noise in the control networks and imperfect maintenance and replication of the epigenome. If such processes do result from increasing disorganization, rather than mere damage, then re-setting epigenetic information using methods derived from those used to produce induced pluripotent cells or to induce transdifferentiation becomes a real possibility to effect rejuvenation and extend life span/health span.

Conclusions

DNA methylation plays a key role in the functioning of the epigenome and maintaining cell differentiation in vertebrates. Changes in DNA methylation are likely to reflect fundamental changes in cell differentiation state. The advent of a widely applicable method to assess biological age in humans directly linked to changes in DNA methylation, reflective of fundamental aging processes at the molecular level, will promote significant advances in anti-aging therapeutic development as well in characterizing how human aging occurs.

References

1. Hannum G, Guinney J, Zhao L, Zhang L, Hughes G, Sadda S, Klotzle B, Bibikova M, Fan J-B, Gao Y, Deconde R, Chen M, Rajapakse I, Friend S, Ideker T, Zhang K. Genome-wide methylation profiles reveal quantitative views of human aging rates. Mol Cell 2013;49:1–9.
2. Masters PM, Bada JL, Zigler JS Jr. Aspartic acid racemisation in the human lens during ageing and in cataract formation. Nature 1977;268:71–73.
3. Truscott RJW. Are ancient proteins responsible for the age-related decline in health and fitness? Rejuvenation Res 2010;13:83–89.
4. Tomas-Loba A, Bernardes de Jesus B, Mayo JM, Blasco MA. A metabolic signature predicts biological age in mice. Aging Cell 2013:12:93–101.
5. Harley CB, Futcher AB, Greider CW. Telomeres shorten during ageing of human fibroblasts. Nature 1990;345: 458–460.
6. Sanders JL, Newman AB. Telomere length in epidemiology: A biomarker of aging, age-related disease, both, or neither? Epidemiol Rev 2013; doi: 10(1)093/epirev/mxs008. Published online: January 9, 2013.
7. Waaijer MEC, Parish WE, Strongitharm BH, Van Heemst D, Slagboom PE, De Craen AJM, Sedivy JM, Westendorp RGJ, Gunn DA, Maier AB. The number of p16ink4a positive cells in human skin reflects biological age. Aging Cell 2012;11: 722–725.
8. Fraser HB, Khaitovich P, Plotkin JB, Paabo S, Eisen MB. Aging and gene

expression in the primate brain. PLoS Biol 2005;3:e274.

9. Zahn JM, Poosala S, Owen AB, Ingram DK, Lustig A, Carter A, Weeraratna AT, Taub DD, Gorospe M, Mazan-Mamczarz K, Lakatta EG, Boheler KR, Xu X, Mattson MP, Falco G, Ko MSH, Schlessinger D, Firman J, Kummerfeld SK, III WHW, Zonderman AB, Kim SK, Becker KG. Agemap: A gene expression database for aging in mice. PLoS Genet. 2007; 3:e201.

10. De Magalha es JP, Curado J, Church GM. Meta-analysis of age-related gene expression profiles identifies common signatures of aging. Bioinformatics 2009;25:875–881.

11. Alisch RS, Barwick BG, Chopra P, Myrick LK, Satten GA, Conneely KN, Warren ST. Age-associated DNA methylation in pediatric populations. Genome Res 2012;22:623–632.

12. Bell JT, Tsai P-C, Yang T-P, Pidsley R, Nisbet J, Glass D, Mangino M, Zhai G, Zhang F, Valdes A, Shin S-Y, Dempster EL, Murray RM, Grundberg E, Hedman AK, Nica A, Small KS, Dermitzakis ET, McCarthy MI, Mill J, Spector TD, Deloukas P. The MuTHER Consortium. Epigenome-wide scans identify differentially methylated regions for age and age-related phenotypes in a healthy ageing population. PLoS Genet 2012;8:e1002629.

13. Bocklandt S, Lin W, Sehl ME, Sanchez FJ, Sinsheimer JS, Horvath S, Vilain E. Epigenetic predictor of age. PLoS One 2011;6:e14821.

14. Boks MP, Derks EM, Weisenberger DJ, Strengman E, Janson E, Sommer IE, Kahn RS, Ophoff RA. The relationship of DNA methylation with age, gender and genotype in twins and healthy controls. PLoS One 2009;4:e6767.

15. Bollati V, Schwartz J, Wright R, Litonjua A, Tarantini L, Suh H, Sparrow D, Vokonas P, Baccarelli A. Decline in genomic DNA methylation through aging in a cohort of elderly subjects. Mech Ageing Dev 2009;130:234–239.

16. Christensen BC, Houseman EA, Marsit CJ, Zheng S, Wrensch MR, Wiemels JL, Nelson HH, Karagas MR, Padbury JF, Bueno R, Sugarbaker DJ, Yeh R-F, Wiencke JK, Kelsey KT. Aging and environmental exposures alter tissue-specific DNA methylation dependent upon CpG island context. PLoS Genet 2009;5:e1000602.

17. Rakyan VK, Down TA, Maslau S, Andrew T, Yang T-P, Beyan H, Whittaker P, McCann OT, Finer S, Valdes AM, Leslie RD, Deloukas P, Spector TD. Human aging-associated DNA hypermethylation occurs preferentially at bivalent chromatin domains. Genome Res 2010;20:434–439.

18. Fraga MF. Genetic and epigenetic regulation of aging. Current Opin Immunol 2009;21:446–453.

19. Martin GM. Epigenetic drift in aging identical twins. Proc Natl Acad Sci USA 2005;102:10413–10414.

20. Vijg J, Campisi J. Puzzles, promises and a cure for ageing. Nature 2008;454:1065–1071.

21. Murgatroyd C, Patchev AV, Wu Y, Micale V, Bockmuhl Y, Fischer D, Holsboer F,

Wotjak CT, Almeida OFX, Spengler D. Dynamic DNA methylation programs persistent adverse effects of early-life stress. Nature Neurosci 2009;12:1559–1566.

22. Guo JU, Ma DK, Mo H, Ball MP, Jang M-H, Bonaguidi MA, Balazer JA, Eaves HL, Xie B, Ford E, Zhang K, Ming G, Gao Y, Song H. Neuronal activity modifies DNA methylation landscape in the adult brain. Nat Neurosci 2011;14:1345–1351.

23. Bibikova M, Barnes B, Tsan C, Ho V, Klotzle B, Le JM, Delano D, Zhang L, Schroth GP, Gunderson KL, Fan J-B, Shen R. High density DNA methylation array with single CpG site resolution. Genomics 2011;98:288–295.

24. Heyn H, Li N, Ferreira HJ, Moran S, Pisano DG, Gomez A, Diez J, Sanchez-Mut JV, Setien F, Carmona FJ, Puca AA, Sayols S, Pujana MA, Serra-Musach J, Iglesias-Platas I, Formiga F, Fernandez AF, Fraga MF, Heath SC, Valencia A, Gut IG, Wang J, Esteller M. Distinct DNA methylomes of newborns and centenarians. Proc Natl Acad Sci USA 2012;109: 10522–10527.

25. Zou, H., and Hastie, T. (2005). Regularization and variable selection via the elastic net. J R Stat Soc Series B Stat Methodol 2005;67:301–320.

26. Feinberg AP, Irizarry RA, Fradin D, Aryee MJ, Murakami P, Aspelund T, Eiriksdottir G, Harris TB, Launer L, Gudnason V, Fallin MD. Personalized epigenomic signatures that are stable over time and co-vary with body mass index. Sci Transl Med 2010;2:49ra67.

27. Bertoni C, Rustagi A, Rando TA. Enhanced gene repair mediated by methyl-CpG-modified single-stranded oligonucleotides Nucleic Acids Res 2009;37:7468–7482.

28. Collins FS, Barker AD. Mapping the cancer genome: Pinpointing the genes involved in cancer will help chart a new course across the complex landscape of human malignancies. Sci Am 2007;296:50–57.

29. Garagnani P, Bacalini MG, Pirazzini C, Gori D, Giuliani C, Mari D, Di Blasio AM, Gentilini D, Vitale G, Collino S, Rezzi S, Castellani G, Capri M, Salvioli S, Franceschi C. Methylation of elovl2 gene as a new epigenetic marker of age. Aging Cell 2012;11:1132–1134.

30. Emilsson V, Thorleifsson G, Zhang B, Leonardson AS, Zink F, Zhu J, Carlson S, Helgason A, Walters GB, Gunnarsdottir S, Mouy M, Steinthorsdottir V, Eiriksdottir GH, Bjornsdottir G, Reynisdottir I, Gudbjartsson D, Helgadottir A, Jonasdottir A, Jonasdottir A, Styrkarsdottir U, Gretarsdottir S, Magnusson KP, Stefansson H, Fossdal R, Kristjansson K, Gislason HG, Stefansson T, Leifsson BG, Thorsteinsdottir U, Lamb JR, Gulcher JR, Reitman ML, Kong A, Schadt EE, Stefansson K. Genetics of gene expression and its effect on disease. Nature 2008;452:423–428.

31. Shannon C, Weaver W. Mathematical Theory of Communication. University of Illinois Press, Urbana IL, 1949.

Chapter 15

Rejuvenation of Adult Stem Cells: Is Age-Associated Dysfunction Epigenetic?

The dysfunctional changes of aging are generally believed to be irreversible due to the accumulation of molecular and cellular damage within an organism's somatic cells and tissues. However, the importance of potentially reversible cell signaling and epigenetic changes in causing dysfunction has not been thoroughly investigated. Striking evidence that increased oxidative stress associated with hematopoietic stem cells (HSCs) from aging mice causes dysfunction has been reported. Forced expression of SIRT3, which activates the reactive oxygen species (ROS) scavenger superoxide dismutase 2 (SOD2) by de-acetylation to reduce oxidative stress, functionally rejuvenates mouse HSCs. These data, combined with numerous other reports, suggest that ROS act as a signal transducer to play a critical regulatory role in HSCs and at least in some other stem cells. It is likely that ectopic expression of SIRT3 restores homeostasis in gene expression networks sensitive to oxidative stress. This result was surprising because age-associated damage from impaired DNA repair had been thought to be irreversible in old HSCs. The effect of up-regulated SIRT3 in HSCs is one of first examples in which intrinsic cellular aging, not apparently associated with changes in the micro-environment, was reversed. However, the stability of rejuvenation in the absence of continued supplemental SIRT3 expression was not investigated. These data are consistent with a hypothesis that potentially reversible processes, such as aberrant signaling and epigenetic drift, are relevant to cellular aging. If true, rejuvenation of at least some aged cells may be simpler than generally appreciated.

Introduction

The dysfunctional changes of aging are generally believed to be irreversible due to the accumulation of molecular and cellular damage within an organism's somatic cells and tissues. Such intrinsic damage has been characterized biochemically, from mutated DNA to oxidized biomolecules to the accumulation of aggregated proteins, etc. (1). The possibility that dysfunction in aging also results from instability of regulatory networks, affecting the epigenetic programming of the differentiated cell state, has been less well explored. Signal transduction, a key component of cell regulatory networks, plays a critical role in maintaining cell state and in regulating gene expression. Aging defects associated with the epigenome have been hypothesized to be reversible (2). Developmental plasticity inherent in the ability to clone organisms from somatic cells, re-program somatic cells into induced pluripotent stem cells (iPSCs) (even from elderly humans, see ref. 3), or trans-differentiate them into cell types as varied as cardiomyocytes or neurons suggest that defects in cellular aging due to regulatory dysfunction may be correctable. In fact, there are well-described examples of rejuvenation of adult stem cell function, but so far each case involves extrinsic changes in the cellular mileau (4). For example, muscle satellite stem cell dysfunction can be restored by blood-borne factors from young animals, forced activation of NOTCH, small hairpin (sh) RNA knockdown of Smad3, or by neutralization of transforming growth factor-beta (TGF-beta) or osteopontin (5–8). However, evidence that intrinsic aging-associated cellular dysfunction may be based in aberrant gene regulation or signal transduction and be potentially reversible by epigenetic reprogramming or tweaking signal transduction is mostly lacking.

One of the most potentially insightful and relevant debates in aging research is whether oxidative stress/redox/ reactive oxygen species (ROS) contribute to aging by causing the accumulation of damage in the mitochondria in a self-perpetuating positive feedback loop, or alternatively by sending dysregulated homeostatic signals (noise). Although many studies have reported a connection between ROS and molecular damage, evidence is accumulating in model organisms, such as nematodes, fruit flies, and even mice, that ROS can alter the rate of aging through signal transduction mechanisms9 and probably plays an important role in control of cell proliferation (10), differentiation, (11–13) metabolism (14), and death (15) For example, endogenous ROS are necessary for proliferation of neural stem and progenitor cells (16) and for tail regeneration in *Xenopus* tadpoles (17). One relatively unexplored possibility is that ROS signal by modifying the gene networks and the epigenome by directly acting on redox-sensitive chromatin remodeling proteins and transcription factors, which might have profound implications for explaining age-correlated epigenetic changes (18). If ROS signaling is as important to homeostasis as recent reports hint, then understanding how ROS levels are controlled is critical to untangling its role in aging. SIRT3, a mammalian member of the sirtuin family of nicotinamide adenine dinucleotide (NAD+)-dependent deacetylases, deacylases, and adenosine diphosphate (ADP)-ribosyltransferases (19) is among the more intriguing candidates for a regulator of ROS (20). Sirtuins have been implicated in aging since it was found that Sir2, a canonical sirtuin, was necessary for increased longevity associated with caloric restriction (CR) in yeast and that over-expression of Sir2 extended life span in a variety of invertebrates. However, the connection between longevity and Sir2 over-expression in worms and flies is controversial (21) and only over-expression of SIRT6 has been found to increase longevity in a mammalian model system and then only male mice (22) Similarly, the role of resveratrol as a sirtuin activator has been debated, although the latest evidence suggests that resveratrol acts directly as an allosteric activator of SIRT1 (23). Nevertheless, sirtuins do play a role in metabolic regulation and of the seven mammalian sirtuins; three—SIRT3, SIRT4, and SIRT5—are mostly localized to the mitochondria, where SIRT3 and SIRT5 help regulate mitochondrial enzyme activities and metabolic adaptation (24) by altering protein acetylation. SIRT3 expression, along with SIRT1 and SIRT6, increases during CR.25–27 SIRT3 affects numerous proteins in electron transport (20) is required for the activity of the mitochondrial biogenesis master regulator peroxisome proliferator-activated receptor-c co-activator (PPARc) in brown adipocytes and skeletal muscle (28,29) modulates cell death in response to injury or genotoxic damage (30, 31) is required for CR to promote long-term survival of inner ear cells involved in hearing that are normally lost during aging (26) and prevents cardiac hypertrophy associated with aging in mice (32, 33) SIRT3 controls ROS levels by stimulating ROS scavengers superoxide dismutase 2 (SOD2) (34,35) and possibly catalase (35) During CR, SIRT3 directly de-acetylates SOD2 to increase superoxide scavenging, (36) while extra-mitochondrial SIRT3 also can de-acetylate the FoxO3A transcription factor, which then trans-locates to the nucleus to increase SOD2 and catalase transcription (35) (but this was not confirmed in other work, see ref. 36). Furthermore, SIRT3 indirectly increases the levels of reduced glutathione, another important cellular anti-oxidant, by increasing isocitrate dehydrogenase 2 activity, which increases nicotinamide adenine dinucleotide phosphate-oxidase (NADPH) levels, which in turn stimulates glutathione reductase (26).

Increased SIRT3 Expression Rejuvenates Aging Hematopoietic Stem Cells

ROS levels increase in hematopoietic stem cells (HSCs) with age, resulting in increased apoptosis and proliferation, which eventually results in diminished self-renewal capacity and differentiation. A positive feedback loop in which ROS result in mitochondrial molecular damage, which in turn causes increased ROS, has been hypothesized to cause irreversible stem cell quiescence, senescence, and death. Aging of HSCs in particular has been thought to be irreversible due to cell-intrinsic, irreversible changes in DNA repair (4, 37, 38). In an important paper, Brown *et al.* (39) show that forced over-expression of SIRT3 in HSCs from old mice restores HSC regenerative capacity. We believe that the simplest explanation is that SIRT3 attenuates dysfunctional mitochondrial ROS signaling, which acts as a source of epigenetic noise to disrupt gene regulatory networks sensitive to redox and oxidative stress, which in turn perturbs networks that control HSC function.

To dissect the role of ROS and SIRT3 in HSC function, Brown and colleagues first tested the hypothesis that loss of SIRT3 increases ROS and reduces function in HSCs. In young mice, SIRT3 is expressed at 3000-fold greater levels in HSCs than in differentiated blood cells, and the authors observed that SIRT3 expression decreases in old animals, so the authors expected that the HSCs of SIRT3 knockout mice (SIRT3-KO) might exhibit premature aging. However, they found no differences in the numbers of HSCs (Lin-, c-Kit+, Sca1+, CD150+, CD48-) or hematopoietic stem/progenitor cells (HSPCs)((Lin-, c-Kit+, Sca1+) from SIRT3-KO mice. (HSCs are a subset of hematopoietic stem and progenitor cells [HSPCs]; Brown and co-workers generally observed similar results for both populations.) HSCs from young SIRT3-KO animals show normal ROS levels and are equally adept as wild type (WT) mice at reconstituting the hematopoietic compartment of lethally irradiated mice. No difference in the number of blood cell types derived from the WT and SIRT3-KO HSCs was observed (39).

ROS levels in HSCs are known to increase significantly as animals age (40) and after three cycles of serial transplantation. HSCs derived from SIRT3-KO mice demonstrated a 50% reduction in self-renewal compared to WT mice after three serial transplantations. Consistent with this observation, HSC compartments were 50% smaller in aged 18-to 24-month-old SIRT3-KO mice compared to WT, and the reconstitution efficiency of aged HSCs was 30% less in for SIRT3-KO cells. The authors concluded that SIRT3 is necessary to maintain HSC pool size and function with increasing age (39). However, there is another important conclusion that the authors do not reach: Because SIRT3 is not necessary to prevent high ROS or dysfunction in young animals, loss of SIRT3 is not the only defect in aging HSCs causing increased ROS. The critical question that remains is the cause of the increased ROS levels with age. The question remains unanswered.

Brown *et al.* studied whether SIRT3-KO affects the microenvironment/niche by transplanting HSCs from WT donors into SIRT3-KO mice. If the absence of SIRT3 altered the stem cell micro-environment, then less self-renewal, reconstitution, and differentiation would be observed in SIRT3-KO mice. No difference in HSC function was seen between irradiated SIRT3-KO and WT mice, suggesting that SIRT3 does not act extrinsically to affect the micro-environment (39).

SIRT3 is known to increase activity of anti-oxidant enzymes such as SOD2 to reduce ROS levels. Although knockout of SIRT3 did not alter ROS levels in HSPCs of young animals, SIRT3-KO did increase ROS in aged HSPCs and under transplant stress. To assess whether the (decreased) presence of SIRT3 still protects old HSPCs from elevated ROS through SOD2, SOD2 expression and activity were compared between old WT and SIRT3-KO HSPCs. HSPCs had equivalent levels of SOD2 mRNAs, but WT mice had 50% more SOD2 activity, whereas SIRT3-KO mice had 40% more dysfunctional mitochondria as assessed by Mitotracker fluorescent dyes. The authors conclude that SIRT3 regulates mitochondrial homeostasis and integrity in HSPCs (39) To further clarify the role of SOD2, old HSCs were infected with lentiviruses that ectopically expressed either WT SOD or with SOD2 carrying K53R and K89R mutations, which prevent acetylation. The mutant is as catalytically active as deacetylated WT SOD2. Interestingly, only forced expression of mutant SOD2 increased colony-forming activity (by 75%); WT SOD2 had no effect. This data is consistent with the hypothesis that SIRT3 acts at least in part by deacetylating SOD2, which then reduces ROS by dismutation of superoxide ions. Normative ROS levels then deactivate oxidative stress–sensitive gene networks that interfere with function. One implication of this result is that old HSCs have high levels of mitochondrial protein acetylation, which may be caused by lower SIRT3 expression.

Oxidative stress disrupts the normal quiescent state that protects HSCs from over-proliferation and subsequent senescence to maintain self-renewal capacity (41). Increased cell cycle activity and decreased survival were observed in aged HSPCs, suggesting that possible disruption of function was due to oxidative stress/ROS. This led the authors to test whether reduction of ROS by other means could rescue old SIRT3-KO cells. Treatment with the powerful anti-oxidant N-acetyl-l-cysteine (NAC) rescued the ability of SIRT3-KO cells in reconstitution experiments, suggesting that reduction of ROS alone can restore stem cell function, and that SIRT3 likely rescues aged cells by reducing ROS through their effects on SOD2 and switching off oxidative stress response.

Most importantly, experiments employing forced overexpression of SIRT3 by transduction with SIRT3 lentiviral vectors similarly reduced ROS and restored regenerative capacity of aged and SIRT3-KO HSCs (39). This is one of first demonstrations that defects associated with intrinsic aging at the cellular level can be reversed in stem cells and that aging cells can be rejuvenated.

Possible Mechanisms for SIRT3 Rejuvenation of HSCs

There are two important questions:

(1) How does ectopic expression of SIRT3 rejuvenate HSCs?

(2) How thoroughly does SIRT3 rejuvenate HSCs?

SIRT3 is clearly not necessary to maintain normative ROS levels and HSC function in young animals; nevertheless, it can rescue HSC function and restore ROS levels in old animals. The authors provide strong evidence that SIRT3 reduces high ROS levels by de-acetylating SOD2 to stimulate removal of superoxide radicals. The resulting absence of high ROS levels acts as signal to terminate oxidative-stress responses in gene networks that inhibit HSC function. One possible mechanism by which ROS may destabilize HSC function is though activation of Nrf2, a transcription factor that acts as an oxidative stress sensor. In the presence of ROS, Nrf2 trans-locates from the cytoplasm to the nucleus to initiate transcription of a protective response at genes regulated by anti-oxidant response elements. In a recent report, Nrf2 was shown to regulate HSC function and renewal (42). We speculate that chronic Nrf2 activation may suppress HSC function, but that hypothesis needs to be investigated.

To answer how thoroughly SIRT3 rejuvenates HSCs requires defining what is necessary for complete rejuvenation. Is restoration of HSC function sufficient or does full rejuvenation require removing additional damaged biomolecules or resetting the cell's epigenetic state such that ROS levels will not immediately return to high levels in the absence of supplemental SIRT3? In other words, it is necessary to understand why ROS levels increased with aging in the first place? One possibility is that increased ROS levels result from accumulating damage that also somehow causes lowered SIRT3, as the authors suggest. Another possibility is that the homeostasis of ROS signaling is disrupted by stochastic fluctuation in signal transduction or perhaps epigenetic drift, such as seen in the human methylome (43-44) In the latter case, a close examination of possible connections between the genes whose DNA methylation state best correlates with age and SIRT3 is warranted. A unifying possibility might be that the use of ROS for cell signaling is itself the source of gene regulatory and epigenetic perturbations, mediated by redox and oxidative stress-sensitive transcription factors and epigenomic regulators.

One way to gain insight into the extent of rejuvenation is to assess whether the effects of chronic oxidative stress are completely removed by the initial treatment. If so, switching off the ectopically expressed SIRT3 or terminating treatment with NAC should result in a gradual return to high ROS and the dysfunctional state over time. If the underlying defect was only masked by by-passing upstream permanent damage or epigenetic information loss, then removing SIRT3 or NAC should result in an immediate return to high ROS and the dysfunctional state.

General Implications for Rejuvenation of Aged Cells

Work on HSCs and satellite skeletal muscle stem cells, for example, suggests that rejuvenation of some cell types is possible. For comprehensive rejuvenation, the true extent of irreversible molecular damage, potentially reversible epigenetic changes, and reversible disruption of cell signaling/ proper cellular milieu must be assessed, so that appropriate techniques to overcome this damage can be developed and applied. For example, the SENS Foundation has cataloged a number of ways in which molecular, cellular, and extracellular damage can be removed and repaired. Obviously, the more plastic the molecular basis of the dysfunction is, potentially the easier it can be addressed. For example, DNA mutations that remain after endogenous DNA repair are essentially irreversible and could only be corrected by the development of safe genome engineering. On the other hand, disruption of gene regulatory networks by aberrant signaling may be treatable by drugs, similar to the effects of NAC on ROS in HSCs in vivo. Epigenetic alterations that stably alter differentiation state, such as those seen in epigenetic drift, lie in a middle ground. It is unclear whether such cells would first require full reprogramming to a pluripotent state followed by re-differentiation, which is completely impractical *in vivo*, or could be directly stimulated to re-differentiate to their own youthful cell state

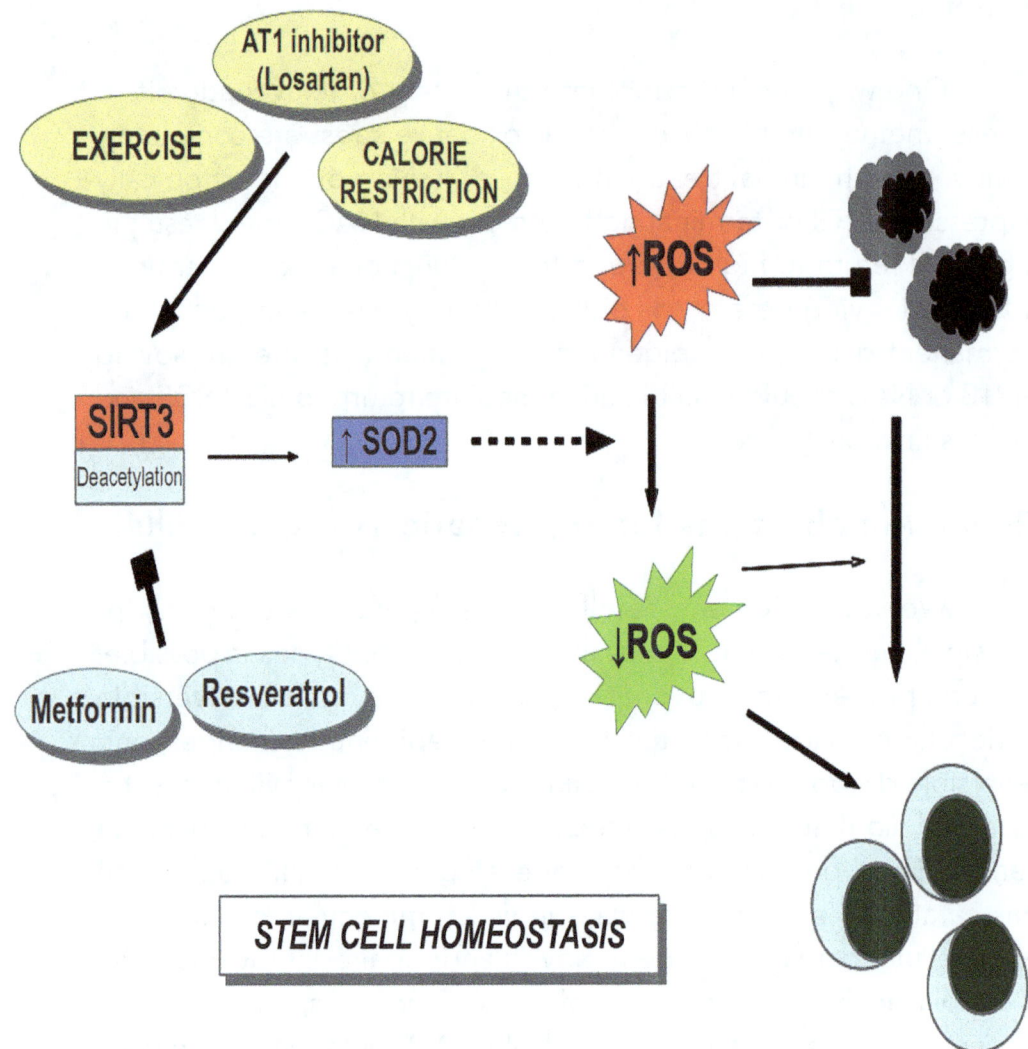

Figure 15.1. **SIRT3 rescues hematopoietic stem cell (HSC) function by reducing reactive oxygen species (ROS).** HSC function is inhibited by high ROS levels in old cells (grey and black). SIRT3 stimulates antioxidant superoxide dismutase 2 (SOD2) activity by deacetylation. SOD2 reduces ROS levels, restoring HSC function (cyan cells). Exercise, calorie restriction (CR), and angiotensin (AT1) inhibitors increase SIRT3 expression. Resveratrol inhibits SIRT3 activity and metformin inhibits SIRT3 expression.

using methods similar to those used to trans-differentiate cells.

Medical Implications

The loss of SIRT3 expression with age or stress demonstrated in HSCs may be more ubiquitous than reported in the work of Brown *et al.* It will be of great interest to assess mitochondrial function, ROS, and SIRT3 levels in other tissues and stem cells to see how widespread this effect is. It will be especially important to determine if SIRT3 is down-regulated with age and whether enforced SIRT3 up-regulation can rescue function in human HSCs or other human stem cells.

Given these intriguing results, are there any known ways to raise SIRT3 expression levels? Three ways have been reported (**Figure 15.1**)—exercise (45), CR (46), and treatment with angiotensin (AT1) antagonists, such as losartan (47). Interestingly, all of these have been reported to extend longevity in at least some animal models.

Just as important is the observation that some compounds thought to extend longevity or to be of benefit in slowing some aging-related processes actually may inhibit SIRT3 expression or activity (**Figure 15.1**). Among these are resveratrol (and some of its metabolites), which at high doses interact with and inhibit SIRT3 enzyme activity (48(, and metformin, which inhibits expression of SIRT3 (49). Perhaps this is one of the reasons why life span extension in mammals is modest at best with metformin and absent in animals fed standard diets supplemented with resveratrol.

Conclusions

The effects of aging are generally thought to be irreversible, due to the accumulation of molecular damage. However, Brown and colleagues show that even presumed intrinsic aging in stem cells may be reversible. Ectopic expression of SIRT3, expression of a constitutively de-acetylated SOD2, and NAC all lower ROS levels, such that HSC function is restored, possibly by terminating oxidative stress-mediated disruption of gene networks involved in regulating HSC function. Whether the HSCs are stably rejuvenated by SIRT3 remains unclear, in terms of whether they retain residual damage or hidden epigenetic changes. The reversibility of aging-associated dysfunction in other somatic cells and stem cells may depend upon the dynamism or plasticity of the underlying molecular changes. If these results are found to hold in humans, the use of AT1 antagonists might have benefit for older people who have reduced SIRT3 expression that leads to immune dysfunction.

References

1. Holliday R. Aging is no longer an unsolved problem in biology. Ann NY Acad Sci 2006;1067:1–9.
2. Sinclair D, Oberdoerffer P. The ageing epigenome: Damaged beyond repair? Ageing Res Rev 2009;8:189–198.
3. Ohmine S, Squillace KA, Hartjes KA, Deeds MC, Armstrong AS, Thatava T, Sakuma T, Terzic A, Kudva Y, Ikeda Y. Reprogrammed keratinocytes from elderly type 2 diabetes patients suppress senescence genes to acquire induced pluripotency. Aging (Albany NY) 2012;4:60–73.
4. Pollina EA, Brunet A. Epigenetic regulation of aging stem cells. Oncogene 2011;30:3105–3126.
5. Paliwal P, Pishesha N, Wijaya D, Conboy IM. Age dependent increase in the levels of osteopontin inhibits skeletal muscle regeneration. Aging (Albany NY) 2012;4:553–566.
6. Carlson ME, Suetta C, Conboy MJ, Aagaard P, Mackey A, Kjaer M, Conboy I. Molecular aging and rejuvenation of human muscle stem cells. EMBO Mol Med 2009;1:381–391.
7. Carlson ME, Hsu M, Conboy IM. Imbalance between PSMAD3 and notch induces CDK inhibitors in old muscle stem cells. Nature 2008;454:528–532.
8. Conboy IM, Conboy MJ, Wagers AJ, Girma ER, Weissman IL, Rando TA. Rejuvenation of aged progenitor cells by exposure to a young systemic environment. Nature 2005;433: 760–764.
9. Back P, Braeckman BP, Matthijssens F. ROS in aging *Caenorhabditis elegans*: Damage or signaling? Oxid Med Cell Longev 2012;2012.
10. Verbon EH, Post JA, Boonstra J. The influence of reactive oxygen species on cell cycle progression in mammalian cells. Gene 2012;511:1–6.
11. Kennedy KAM, Sandiford SDE, Skerjanc IS, Li SS-C. Reactive oxygen species

and the neuronal fate. Cell Mol Life Sci 2012;69:215–221.

12. Sardina JL, Lopez-Ruano G, Sanchez-Sanchez B, Llanillo M, Hernandez-Hernandez A. Reactive oxygen species: Are they important for haematopoiesis? Crit Rev Oncol Hematol 2012;81:257–274.

13. Schippers JHM, Nguyen HM, Lu D, Schmidt R, Mueller-Roeber B. ROS homeostasis during development: An evolutionary conserved strategy. Cell Mol Life Sci 2012;69: 3245–3257.

14. Page MM, Robb EL, Salway KD, Stuart JA. Mitochondrial redox metabolism: Aging, longevity and dietary effects. Mech Ageing Dev 2010;131:242–252.

15. Apostolova N, Blas-Garcia A, Esplugues JV. Mitochondria sentencing about cellular life and death: A matter of oxidative stress. Curr Pharm Des 2011;17:4047–4060.

16. Yoneyama M, Kawada K, Gotoh Y, Shiba T, Ogita K. Endogenous reactive oxygen species are essential for proliferation of neural stem/progenitor cells. Neurochem Int 2010; 56:740–746.

17. Love NR, Chen Y, Ishibashi S, Kritsiligkou P, Lea R, Koh Y, Gallop JL, Dorey K, Amaya E. Amputation-induced reactive oxygen species are required for successful Xenopus tadpole tail regeneration. Nat Cell Biol 2012;15:222–228.

18. Cyr AR, Domann FE. The redox basis of epigenetic modifications: From mechanisms to functional consequences. Antioxid Redox Signal 2011;15:551–589.

19. Sebastian C, Satterstrom FK, Haigis MC, Mostoslavsky R. From sirtuin biology to human diseases: An update. J Biol Chem 2012;287:42444–42452.

20. Sack MN, Finkel T. Mitochondrial metabolism, sirtuins, and aging. Cold Spring Harb Perspect Biol 2012;4.

21. Burnett C, Valentini S, Cabreiro F, Goss M, Somogyvari M, Piper MD, Hoddinott M, Sutphin GL, Leko V, McElwee JJ, Vazquez-Manrique RP, Orfila A-M, Ackerman D, Au C, Vinti G, Riesen M, Howard K, Neri C, Bedalov A, Kaeberlein M, Soti C, Partridge L, Gems D. Absence of effects of SIR2 overexpression on lifespan in *C. elegans* and Drosophila. Nature 2011;477:482–485.

22. Kanfi Y, Naiman S, Amir G, Peshti V, Zinman G, Nahum L, Bar-Joseph Z, Cohen HY. The sirtuin SIRT6 regulates lifespan in male mice. Nature 2012;483:218–221.

23. Hubbard BP, Gomes AP, Dai H, Li J, Case AW, Considine T, Riera TV, Lee JE, E SY, Lamming DW, Pentelute BL, Schuman ER, Stevens LA, Ling AJY, Armour SM, Michan S, Zhao H, Jiang Y, Sweitzer SM, Blum CA, Disch JS, Ng PY, Howitz KT, Rolo AP, Hamuro Y, Moss J, Perni RB, Ellis JL, Vlasuk GP, Sinclair DA. Evidence for a common mechanism of SIRT1 regulation by allosteric activators. Science 2013;339:1216–1219.

24. Lombard DB, Tishkoff DX, Bao J. Mitochondrial sirtuins in the regulation of mitochondrial activity and metabolic adaptation. Handb Exp Pharmacol 2011;206:163–188.

25. Cohen HY, Miller C, Bitterman KJ, Wall NR, Hekking B, Kessler B, Howitz KT, Gorospe M, De Cabo R, Sinclair DA. Calorie restriction promotes mammalian cell

survival by inducing the SIRT1 deacetylase. Science 2004;305:390–392.
26. Someya S, Yu W, Hallows WC, Xu J, Vann JM, Leeuwenburgh C, Tanokura M, Denu JM, Prolla TA. SIRT3 mediates reduction of oxidative damage and prevention of age-related hearing loss under caloric restriction. Cell 2010;143:802–812.
27. Kanfi Y, Shalman R, Peshti V, Pilosof SN, Gozlan YM, Pearson KJ, Lerrer B, Moazed D, Marine J-C, De Cabo R, Cohen HY. Regulation of SIRT6 protein levels by nutrient availability. FEBS Lett 2008;582:543–548.
28. Kong X, Wang R, Xue Y, Liu X, Zhang H, Chen Y, Fang F, Chang Y. Sirtuin 3, a new target of PGC-1a, plays an important role in the suppression of ROS and mitochondrial biogenesis. PLoS One 2010;5:e11707.
29. Giralt A, Hondares E, Villena JA, Ribas F, Dıaz-Delfın J, Giralt M, Iglesias R, Villarroya F. Peroxisome proliferator activated receptor-c coactivator-1a controls transcription of the SIRT3 gene, an essential component of the thermogenic brown adipocyte phenotype. J Biol Chem 2011;286:16958– 16966.
30. Yang H, Yang T, Baur JA, Perez E, Matsui T, Carmona JJ, Lamming DW, Souza-Pinto NC, Bohr VA, Rosenzweig A, De Cabo R, Sauve AA, Sinclair DA. Nutrient-sensitive mitochondrial NAD + levels dictate cell survival. Cell 2007;130: 1095–1107.
31. Sundaresan NR, Samant SA, Pillai VB, Rajamohan SB, Gupta MP. SIRT3 is a stress-responsive deacetylase in cardiomyocytes that protects cells from stress-mediated cell death by deacetylation of Ku70. Mol Cell Biol 2008;28:6384– 6401.
32. Sack MN. The role of SIRT3 in mitochondrial homeostasis and cardiac adaptation to hypertrophy and aging. J Mol Cell Cardiol 2012;52:520–525.
33. Pillai VB, Sundaresan NR, Kim G, Gupta M, Rajamohan SB, Pillai JB, Samant S, Ravindra PV, Isbatan A, Gupta MP. Exogenous NAD blocks cardiac hypertrophic response via activation of the SIRT3-LKB1-AMP-activated kinase pathway. J Biol Chem 2010;285:3133–3144.
34. Qiu X, Brown K, Hirschey MD, Verdin E, Chen D. Calorie restriction reduces oxidative stress by SIRT3-mediated SOD2 activation. Cell Metab. 2010;12:662–667.
35. Sundaresan NR, Gupta M, Kim G, Rajamohan SB, Isbatan A, Gupta MP. Sirt3 blocks the cardiac hypertrophic response by augmenting Foxo3a-dependent antioxidant defense mechanisms in mice. J Clin Invest 2009;119:2758–2771.
36. Tao R, Coleman MC, Pennington D, Ozden O, Park S-H, Jiang H, Kim H-S, Flynn CR, Hill S, McDonald WH, Olivier AK, Spitz DR, Gius D. Sirt3-mediated deacetylation of evolutionarily conserved lysine 122 regulates MnSOD activity in response to stress. Mol Cell 2010;40:893–904.
37. Nijnik A, Woodbine L, Marchetti C, Dawson S, Lambe T, Liu C, Rodrigues NP, Crockford TL, Cabuy E, Vindigni A, Enver T, Bell JI, Slijepcevic P, Goodnow CC, Jeggo PA, Cornall RJ. DNA repair is limiting for haematopoietic stem cells during ageing. Nature 2007;447:686–690.

38. Rossi DJ, Bryder D, Seita J, Nussenzweig A, Hoeijmakers J, Weissman IL. Deficiencies in DNA damage repair limit the function of haematopoietic stem cells with age. Nature 2007;447:725–729.

39. Brown K, Xie S, Qiu X, Mohrin M, Shin J, Liu Y, Zhang D, Scadden DT, Chen D. SIRT3 reverses aging-associated degeneration. Cell Rep 2013;3:319–327.

40. Ito K, Hirao A, Arai F, Takubo K, Matsuoka S, Miyamoto K, Ohmura M, Naka K, Hosokawa K, Ikeda Y, Suda T. Reactive oxygen species act through p38 MAPK to limit the lifespan of hematopoietic stem cells. Nat Med 2006;12:446–451.

41. Rossi DJ, Jamieson CHM, Weissman IL. Stems cells and the pathways to aging and cancer. Cell 2008;132:681–696.

42. Tsai JJ, Dudakov JA, Takahashi K, Shieh J-H, Velardi E, Holland AM, Singer NV, West ML, Smith OM, Young LF, Shono Y, Ghosh A, Hanash AM, Tran HT, Moore MAS, Van den Brink MRM. Nrf2 regulates haematopoietic stem cell function. Nat Cell Biol 2013;15:309–316.

43. Hannum G, Guinney J, Zhao L, Zhang L, Hughes G, Sadda S, Klotzle B, Bibikova M, Fan J-B, Gao Y, Deconde R, Chen M, Rajapakse I, Friend S, Ideker T, Zhang K. Genome-wide methylation profiles reveal quantitative views of human aging rates. Mol Cell 2013;49:359–367.

44. Mendelsohn AR, Larrick JW. The DNA methylome as a biomarker for epigenetic instability and human aging. Rejuvenation Res 2013;16:74–77.

45. Palacios OM, Carmona JJ, Michan S, Chen KY, Manabe Y, III JLW, Goodyear LJ, Tong Q. Diet and exercise signals regulate SIRT3 and activate AMPK and PGC-1a in skeletal muscle. Aging (Albany NY) 2009;1:771–783.

46. Hebert AS, Dittenhafer-Reed KE, Yu W, Bailey DJ, Selen ES, Boersma MD, Carson JJ, Tonelli M, Balloon AJ, Higbee AJ, Westphall MS, Pagliarini DJ, Prolla TA, Assadi-Porter F, Roy S, Denu JM, Coon JJ. Calorie restriction and SIRT3 trigger global reprogramming of the mitochondrial protein acetylome. Mol Cell 2013;49:186–199.

47. Benigni A, Corna D, Zoja C, Sonzogni A, Latini R, Salio M, Conti S, Rottoli D, Longaretti L, Cassis P, Morigi M, Coffman TM, Remuzzi G. Disruption of the Ang II type 1 receptor promotes longevity in mice. J Clin Invest 2009;119: 524–530.

48. Gertz M, Nguyen GTT, Fischer F, Suenkel B, Schlicker C, Franzel B, Tomaschewski J, Aladini F, Becker C, Wolters D, Steegborn C. A molecular mechanism for direct sirtuin activation by resveratrol. PloS One 2012;7:e49761.

49. Buler M, Aatsinki S-M, Izzi V, Hakkola J. Metformin reduces hepatic expression of SIRT3, the mitochondrial deacetylase controlling energy metabolism. PloS One 2012;7.

Chapter 16

Rejuvenation of aging hearts

Introduction

Specific, subtle changes in regulation or activity of factors that maintain homeostasis and cell differentiation may play significant roles in mammalian aging. Both extrinsic and intrinsic reversible aging-associated changes have been described that can profoundly influence tissue function. There are numerous examples of aging-associated extrinsic changes to the extracellular milieu that affect cell function. For example, muscle satellite stem cell dysfunction correlates with loss of NOTCH function and can be restored by blood- borne factors from young animals, forced activation of NOTCH, small hairpin RNA (sh RNA) knockdown of Smad3, or by neutralization of transforming growth factor-beta (TGF-beta) or osteopontin (1–5).

An example of an intrinsic change is the loss of function in hematopoietic stem cells (HSCs) from old mice. Dysfunction in HSCs correlates with increased reactive oxygen species (ROS) and the down-regulation of sirtuin-3 (SIRT3) in old HSCs. Forced expression of SIRT3 is sufficient to restore HSC function (6,7). Whether due to instability in epigenetic-based cell differentiation programs, such as that hypothesized to result from developmental drift, or to accumulation of damage in critical cell populations, the amplification of small differences in levels or activity of key regulators may play a profound role in aging and provide new targets for therapeutic intervention.

Parabiosis Between Old and Young Mice or Growth Differentiation Factor 11 Treatment Rejuvenates Old Mouse Hearts

Diastolic dysfunction in elderly mice and humans often leads to diastolic heart failure and death. In old mice, as well as long-lived humans, heart failure is one of the most frequent direct causes of death, thus a treatment with potential to rejuvenate heart function may prove helpful in extending longevity. In an important study with significant ramifications for mammalian aging, Loffredo et al. (8) report that it is possible in old mice to reverse cardiac hypertrophy that may underlie age-associated ventricular stiffening and diastolic dysfunction (9). Loffredo et al. (8) hypothesized that extrinsic circulating factors in young mice might reverse cardiac aging. Implicit in this hypothesis is that maladaptive changes in circulating factors occur during aging. Evidence for systemic factors playing a profound role in aging has been found previously for skeletal muscle regeneration1 and neurogenesis (10). Reversibility of cardiac aging in mice previously has been reported by an intrinsic intervention— the overexpression of catalase (11). As in the studies on skeletal muscle and neurogenesis, the effects of surgically linking the circulatory systems ("parabiosis") of an old and young mouse ("heterochronic parabiosis") are compared with controls in which an old mouse is linked to an old mouse or a young mouse to a young mouse ("isochronic parabiosis"). In this study, male and female C57BL/6 mice were used because they spontaneously develop cardiac hypertrophy with age (11). After 4 weeks of cross circulation, hearts from old mice paired with young mice were significantly smaller than those from old mice paired with old mice. Furthermore, hearts from the old partner of old/young mouse pairs weighed less than controls when normalized to tibia length (a standard way to control for size differences in mice). Histological studies revealed that cardiac myocyte cross-sectional area was restored to that of young mice or paired young mice. At the molecular level, transcription of cardiac stress–associated brain natriuretic peptide (BNP) and atrial natriuretic peptide (ANP) were reduced in the hearts of old mice paired with young mice, whereas levels of sarcoplasmic-endoplasmic reticulum Ca++ATPase (SERCA)-2 mRNA, which is associated with diastolic relaxation, were increased. Altogether, these data suggested that 4 weeks of parabiosis was sufficient to reduce cardiac hypertrophy to youthful levels in old mice. Interestingly, the hearts of the young mice that had been paired with old mice did not show any signs of hypertrophy in these experiments. These results suggest that young mice likely synthesize an anti-hypertrophic factor or factors that can restore age-associated diastolic heart failure in old mice, but that old mice do not synthesize a factor that promotes hypertrophy, since the young hearts were not altered by parabiosis with old animals (8).

Specific subtle changes in regulation or activity of factors that maintain homeostasis and cell differentiation may play significant roles in mammalian aging. Drift resulting from reaching the end of an organism's developmental program might involve a specific ordered set of changes. Several studies have suggested that dysfunctional changes associated with aging in skeletal muscle, neurons, and hematopoietic stem cells may be caused by specific changes either in the extracellular environment or in intracellular regulatory networks and that such dysfunction may be reversible. On the basis these data, Loffredo *et al.* hypothesized that extrinsic circulating factors in young mice might reverse cardiac aging. Parabiosis, the surgical linking of circulations between old and young mice, was employed to identify an anti-hypertrophic factor (growth differentiation factor 11 [GDF-11]) that appears to rejuvenate aging murine hearts, raising exciting prospects for the development of anti-aging therapeutics. However, much work remains to be done to evaluate the utility of GDF-11 as a therapeutic rejuvenation factor. Similar rejuvenating factors for diverse tissues may exist as well and will hopefully be identified in the near future.

Because of the possibility of artifactual results due to indirect physical effects, several control studies were performed. The reduced cardiac hypertrophy was not caused by reduced blood pressure or lower heart rate. Moreover circulating levels of angiotensin II and aldosterone, which can affect blood pressure, were not affected by parabiosis. Effects due to altered physical activity or behavior resulting from the surgical connection of the mice was ruled out in sham parabiosis experiments in which mice were sutured together in a similar way to the actual parabiosed mice, but with no connection made between the circulatory systems (8).

Because the parabiosis experiments suggested the existence of an anti-hypertrophic factor or factors synthesized by young mice, an attempt was made to identify candidate factors. Metabolomic profiling of 69 amino acids and amines revealed no differences between plasma from heterochronic or isochronic parabiosed mice. Similarly, lipidomic analysis of 142 lipids from nine lipid classes revealed no differences. However, a proteomic analysis using RNA aptamers that specifically bind over *1100 targets revealed 13 peptides that distinguish young mice from old mice. More extensive analysis of plasma for levels of each of these analytes identified growth differentiation factor 11 (GDF-11) as having the expected expression pattern for an anti-hypertrophic factor. GDF-11 levels were higher in young mice, lower in old mice, and increased in old mice paired with young mice. GDF-11 expression is widespread in mouse tissues, but highest in the spleen. Old mice express much less GDF-11 in the spleen, compared with young mice, consistent with the observed plasma levels (8).

In cultured rat neonatal cardiomyocytes treated with phenylephrine to stimulate hypertrophy, which is measured by increases in [3H]leucine incorporation, GDF-11 inhibits the hypertrophic effects of phenylephrine. Interestingly, myostatin, a protein very closely related structurally to GDF-11, which inhibits skeletal muscle hypertrophy, has no effect on phenylephrine-stimulated [3H]leucine incorporation, indicating that GDF-11 may exert a specialized antihypertrophic effect on cardiomyocytes. In a preliminary experiment to assess how GDF-11 may differ from myostatin in signal transduction, human induced pluripotent stem cell (iPSC)-derived cardiomyocytes were treated with either GDF-11 or myostatin. However, no differences in activation of downstream pSMAD2, pSMAD3, or suppression of Forkhead transcription factor were observed, suggesting uncharacterized signaling differences between the two factors are yet to be elucidated (8).

Daily intra-peritoneal injection of GDF-11 for 30 days into old mice resulted in many of the same effects seen in the heterochronic parabiosis experiments. Treated mouse hearts were smaller and weighed less (after tibia standardization; see above), and cross area was reduced compared to untreated controls. Transcription levels of ANP and BNP were again reduced, and conversely SERCA-2 levels were increased. It is important to note that the magnitude of these effects was not nearly as strong as that seen in the parabiosis experiments, suggesting a contribution by other antihypertrophic factors. The effects of GDF-11 were observed to be specific to age-associated hypertrophy and not pressure overload–induced hypertrophy: In young mice subjected to aortic constriction, 30 days of treatment with GDF-11 had no effect on hypertrophy (8).

Altogether, these results suggest that dysregulation of anti-hypertrophic factors such as GDF-11 may play an important role in cardiac aging. However, these results are only preliminary. It will be essential that these studies be extended to assess actual changes in cardiac function, as well as repeated in other model systems by other researchers.

Medical Implications

Anti-hypertrophic factors such as GDF-11 may be candidates

for therapeutics to prevent or reverse cardiac aging. Reversal of cardiac aging may extend longevity, as was reported for by Dai *et al.* for the overexpression of catalase in mouse mitochondria (11). This may prove of significant benefit in elderly populations. However, the need for replication in animal models and, if confirmed, subsequent human clinical trials highlights the preliminary status of GDF-11 as a rejuvenation agent.

One of the potential problems with GDF-11 is that it may play significant roles in other tissues. Knockout mice lacking GDF-11 die at birth, with skeletal and renal abnormalities, suggesting that GDF-11's role in these tissues should be examined. GDF-11 may also play a role in pancreatic b-cell function (12,13). Furthermore, GDF-11 has been observed to inhibit neurogenesis (14), a potential unwanted side effect. These pitfalls suggest that modulation of the diverse activities of GDF-11 may be required for development of effective therapeutics.

The identification of drugs or nutraceuticals that stimulate native GDF-11 expression may have some utility and should be pursued. For example, it has been reported that the histone deacetylase inhibitor trichostatin A stimulates transcriptional expression of GDF-11 (15). Although trichostatin A has broad physiological effects, and is probably not suitable for inducing GDF-11 clinically, there is reasonable hope that more specific drugs that stimulate GDF-11 can be identified.

Conclusions

Identification of an anti-hypertrophic factor such as GDF-11 that appears to rejuvenate aging murine hearts raises exciting prospects for the development of anti-aging therapeutics. Similar rejuvenating factors for diverse tissues are expected to exist as well and will hopefully be identified in the near future. However, much work remains to be done to evaluate the utility of GDF-11 as a therapeutic rejuvenation factor.

References

1. Conboy IM, Conboy MJ, Wagers AJ, Girma ER, Weissman IL, Rando TA. Rejuvenation of aged progenitor cells by exposure to a young systemic environment. Nature 2005;433: 760–764.
2. Carlson ME, Hsu M, Conboy IM. Imbalance between pSmad3 and Notch induces CDK inhibitors in old muscle stem cells. Nature 2008;454:528–532.
3. Carlson ME, Suetta C, Conboy MJ, Aagaard P, Mackey A, Kjaer M, et al. Molecular aging and rejuvenation of human muscle stem cells. EMBO Mol Med 2009;1:381–391.
4. Carlson ME, Conboy MJ, Hsu M, Barchas L, Jeong J, Agrawal A, et al. Relative roles of TGF-b1 and Wnt in the systemic regulation and aging of satellite cell responses. Aging Cell 2009;8:676–689.
5. Paliwal P, Pishesha N, Wijaya D, Conboy IM. Age dependent increase in the levels of osteopontin inhibits skeletal muscle regeneration. Aging 2012;4:553–566.
6. Brown K, Xie S, Qiu X, Mohrin M, Shin J, Liu Y, et al. SIRT3 reverses aging-associated degeneration. Cell Rep 2013;3:319–327.
7. Mendelsohn AR, Larrick JW. Rejuvenation of adult stem cells: Is age-associated dysfunction epigenetic? Rejuvenation Res 2013;16:152–7.
8. Loffredo FS, Steinhauser ML, Jay SM, Gannon J, Pancoast JR, Yalamanchi P, et al. Growth differentiation factor 11 is a circulating factor that reverses age-related cardiac hypertrophy. Cell 2013;153:828–839.
9. Aurigemma GP. Diastolic heart failure—a common and lethal condition by any name. N Engl J Med 2006;355:308–310.
10. Villeda SA, Luo J, Mosher KI, Zou B, Britschgi M, Bieri G, et al. The ageing systemic milieu negatively regulates neurogenesis and cognitive function. Nature 2011;477:90–94.
11. Dai D-F, Santana LF, Vermulst M, Tomazela DM, Emond MJ, MacCoss MJ, et al. Overexpression of catalase targeted to mitochondria attenuates murine cardiac aging. Circulation 2009;119:2789–2797.
12. Dichmann DS, Yassin H, Serup P. Analysis of pancreatic endocrine development in GDF11-deficient mice. Dev Dyn Off Publ Am Assoc Anat 2006;235:3016–3025.
13. Harmon EB, Apelqvist AA, Smart NG, Gu X, Osborne DH, Kim SK. GDF11 modulates NGN3+ islet progenitor cell number and promotes b-cell differentiation in pancreas development. Development 2004;131:6163–6174.
14. Wu H-H, Ivkovic S, Murray RC, Jaramillo S, Lyons KM, Johnson JE, et al. Auto-

regulation of neurogenesis by GDF11. Neuron 2003;37:197–207.

15. Zhang X, Wharton W, Yuan Z, Tsai S-C, Olashaw N, Seto E. Activation of the growth-differentiation factor 11 gene by the histone deacetylase (HDAC) inhibitor trichostatin A and repression by HDAC3. Mol Cell Biol 2004;24:5106–5118.

Appendix

APPLIED HEALTHSPAN ENGINEERING

According to the Homeric Hymn to Aphrodite, when Eos asked Zeus for Tithonus to be granted immortality, she forgot to ask for eternal youth. Applied Healthspan Engineering (AHE) seeks to address this problem. All organisms have a minimal level of functional reserve required to sustain life that eventually declines to a point incompatible with survival at death. AHE seeks to maintain or restore optimal functional reserve of critical tissues and organs. Tissue reserve correlates with well being. Diet, physical exercise, and currently available small-molecule-based therapeutics may attenuate the rate of decline of specific organs or organ systems, but are unlikely to restore lost reserve. Inherent evolutionary-derived limitations in tissue homeostasis and cell maintenance necessitate the development of therapies to enhance regenerative processes and possibly replace whole organs or tissues. AHE supports the study of cell, tissue, and organ homeostatic mechanisms to derive new regenerative and tissue replacement therapies to extend the period of human health.

Introduction

Tithonus indeed lived forever:

"but when loathsome old age pressed full upon him, and he could not move nor lift his limbs, this seemed to her in her heart the best counsel: she laid him in a room and put to the shining doors. There he babbles endlessly, and no more has strength at all, such as once he had in his supple limbs (Homeric Hymn to Aphrodite)." (1).

What is Applied Healthspan Engineering?

All organisms have a minimal level of functional reserve

required to sustain life that declines to a point incompatible with survival at death. Applied Healthspan Engineering (AHE) seeks to maintain or restore optimal functional reserve of critical tissues and organs. A minimal outcome of successful AHE will be to "square" the human healthspan curve (**Figure A.1**). If this approach is successful, more people will live to a healthy "old age" with a shorter period of morbidity before death. We anticipate that the median lifespan will increase, although increases in maximum lifespan are less predictable.

Most multicellular organisms follow a familiar trajectory, the cycle of life. Young individuals grow to maturity, reproduce, and subsequently age and die. To assure successful reproduction, by its inherent mechanism natural selection can only maximize well being during the reproductive years. After cessation of reproduction, the body declines in function and survives only as long as sufficient physiological reserve is maintained. Even in some single-cell species (e.g., yeast), maximization of healthspan appears to correlate with reproductive fitness. For example, mother cells selectively accumulate damaged proteins, permitting more vital daughter cells a better chance of survival and reproduction (2). Thus, inherent limitations to longevity and ultimately healthspan have been incorporated into the blueprints of each organism by natural selection. AHE seeks to overcome the inherent limitations introduced into the construction of the human body and its tissue and organs. We hypothesize that strategies to augment physiological reserve will increase the healthspan.

The human healthspan begins at conception, with many antecedent, parental, in utero influences (3, 4) and continues until death (see **Figure A.1**). The first two decades of life are characterized by growth and development. Maximum vitality is attained in the late teens and twenties, followed by gradual degeneration, accumulating morbidity and eventually death. In the United States, median life expectancy has increased from about 45

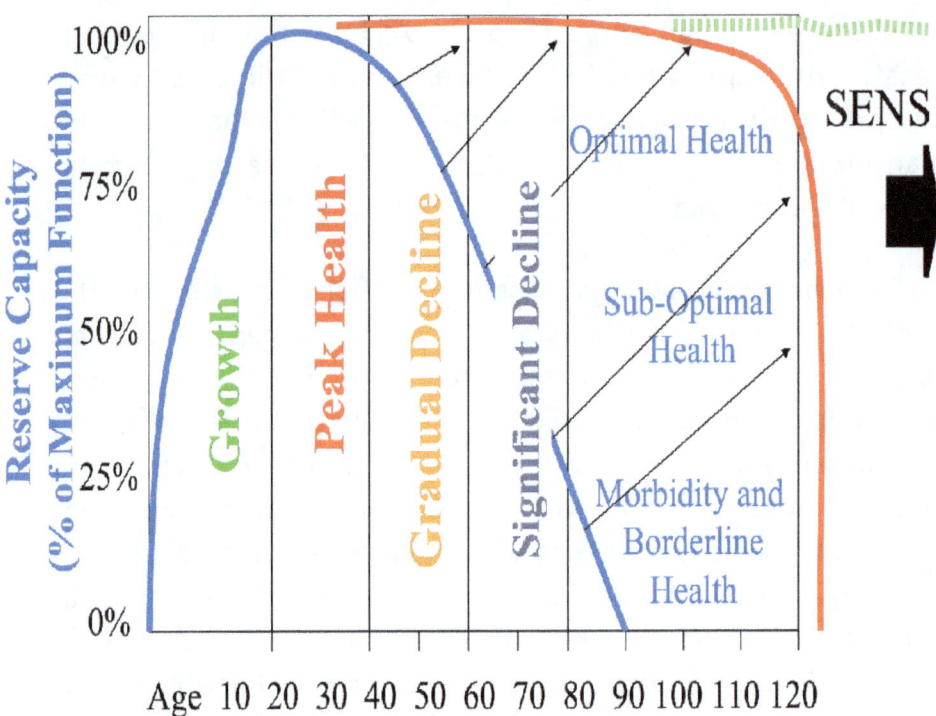

Figure A.1 Minimal goal of Applied Healthspan Engineering (AHE) is to square the healthspan curve. Well-being peaks at the end of growth phase and then declines. AHE seeks to stabilize well-being, possibly resulting in increased longevity. Strategies for Engineered Negligible Senescence (SENS) seeks to increase longevity by "defeating" aging (5).

years in 1900 to 80+ years in 2010.

Graying of the baby boomers has created a large population of so-called "zoomers," people in relatively good health over the age of 50 (www.demko.com/zoomers.htm). A convenient model to consider and compare interventions employs the concept of functional reserve capacity. Although somewhat arbitrary (in the sense that optimal performance for an athlete might greatly exceed reserve capacity of a healthy sedentary person), optimal health corresponds to perhaps 75–100% of maximal function. At <30% of maximal function, a person is likely in suboptimal health. Obviously this is a generalized schema wherein a given tissue or organ system may be limiting when the remainder of the body's functional reserve is adequate.

AHE will use behavioral, pharmacological, biomechanical, and regenerative means to maximize physiological reserve and wellness at increasing chronological age. The initial focus is to modulate pre-existing homeostatic mechanisms, such as blood pressure, heart rate, energy metabolism, etc. When AHE reaches hard limits (those that require genetic or cellular modification) imposed by natural selection, it will be necessary to switch emphasis to engineer new regenerative capacity. Like traditional disease-focused aging research, AHE will promote advances in medical treatments specific to organ systems and tissues. In contrast, traditional molecular aging research seeks to elucidate universal biochemical mechanisms of longevity that might be used to increase human lifespan. By focusing on maximizing organ/tissue functional reserve instead of longevity, AHE represents a practical, incremental alternative that will also likely increase longevity. However, AHE-mediated improvements in tissue reserve, especially in organs such as muscle or skin that are not usually the first point of failure, will not necessarily lead to increased lifespan. Furthermore, although AHE may provide increased well-being, decline due to cancer or accident may lead to rapid death (**Figure A.1**).

AHE has much in common with Strategies for Engineered Negligible Senescence (SENS), which seeks to "defeat aging" by repairing tissue damage.5 The essential difference is that AHE seeks to maintain tissue homeostasis first by modulating existing homeostatic and regenerative mechanisms and then in the future by re-engineering the regenerative capacity of the tissue at the cellular level. SENS seeks to create new means to repair the cellular damage that underlies tissue dysfunction, including direct rejuvenation of aging differentiated cells. Importantly, AHE emphasizes what is available presently as well as future innovations.

What Is Possible? The Upper Limits to Mammalian Healthspan and Longevity

Humans are among the longest-living mammals (6). Although mice usually live less than 4 years, the naked mole rat lives >28 years and similar-sized individuals with very high metabolic rates such as bats have survived >35 years. Dogs live up to 21 years, cats as long as 36 years, and horses up to 62 years (6) A French woman, supercentenarian Jeanne Calment, lived for 122 years, 164 days, the longest documented human lifespan. Among mammals, whales have the longest documented lifespans. Prof. Jeffrey Bada (Scripps Research Institute, La Jolla, CA) used the amount of racemized aspartic acid (L-asp-->R-asp) obtained from the lens of a bowhead whale to estimate its age as 211 years +/-16% (7). Perhaps AHE can benefit from a better understanding of the metabolic basis for the long lives of these creatures with whom we share common ancestors. Consistent with recommendations of most healthcare professionals, whales abstain from smoking, live a low-stress lifestyle and exercise daily. In addition, cetaceans consume fewer glycotoxic and lipotoxic substances than is found in the typical Western diet!

Whales have the slowest heart rates among mammals, which supports a remarkable correlation between heart rate (a reflection of autonomic nervous system activity) and lifespan (8,9) Mean life expectancy among mammals correlates with total lifetime heartbeats (**Figure A.2**). Perhaps this simple relationship instructs AHE to employ vagal stimulation or other means to reduce heart rate to maintain cardiopulmonary reserve (10). In mammals, degenerative processes affecting the cardiovascular system with eventual pathological changes in parenchymal and the extracellular matrix-fibroconnective tissue takes place in virtually all organ systems (11). For example, both quality of life and viability are linked to deterioration of the cardiovascular, pulmonary, and, less commonly, renal systems. In the United States, heart failure is the most common cause of hospitalization, and 93% of deaths in adults over age 55 (12) and among centenarians, by definition individuals who have maximized lifetime fitness, result from failure of cardiopulmonary reserve (68% cardiovascular, 25% pulmonary)(13). Hence, if AHE can optimize cardiopulmonary functional reserve to permit a majority of the population to attain supercentenarian health, it will be considered to be a success.

AHE Seeks to Identify Interventions That Augment Physiological Reserve: A Framework for Discussion of Healthspan Engineering

Wellness diminishes with age. Advancing age correlates with an increasing probability of lost vitality and function in all organ systems. The reserve and vitality of each tissue can be represented by a two-dimensional box the area of which shrinks over time (**Figure A.3**). At some future time, physiological reserve has diminished to a level incompatible with survival. Alternatively, an insult earlier in life can push one "outside the box" of physiological reserve. This can be lethal if the tissue is not returned to homeostasis within a reasonable period of time. For example, until the 1940s many people died of infectious diseases that are today routinely treated with widely available antibiotics. The potentially lethal streptococcal infection that affected Ann Miller in 1942 is a classic example of this model. At age 30, Ms. Miller sustained a life-threatening throat infection which left her deleterious and febrile to 107F. Luckily, she was the first patient to receive penicillin, which saved her life(14). By pushing her "back into the box," penicillin permitted her to live an additional 60 years! The success of

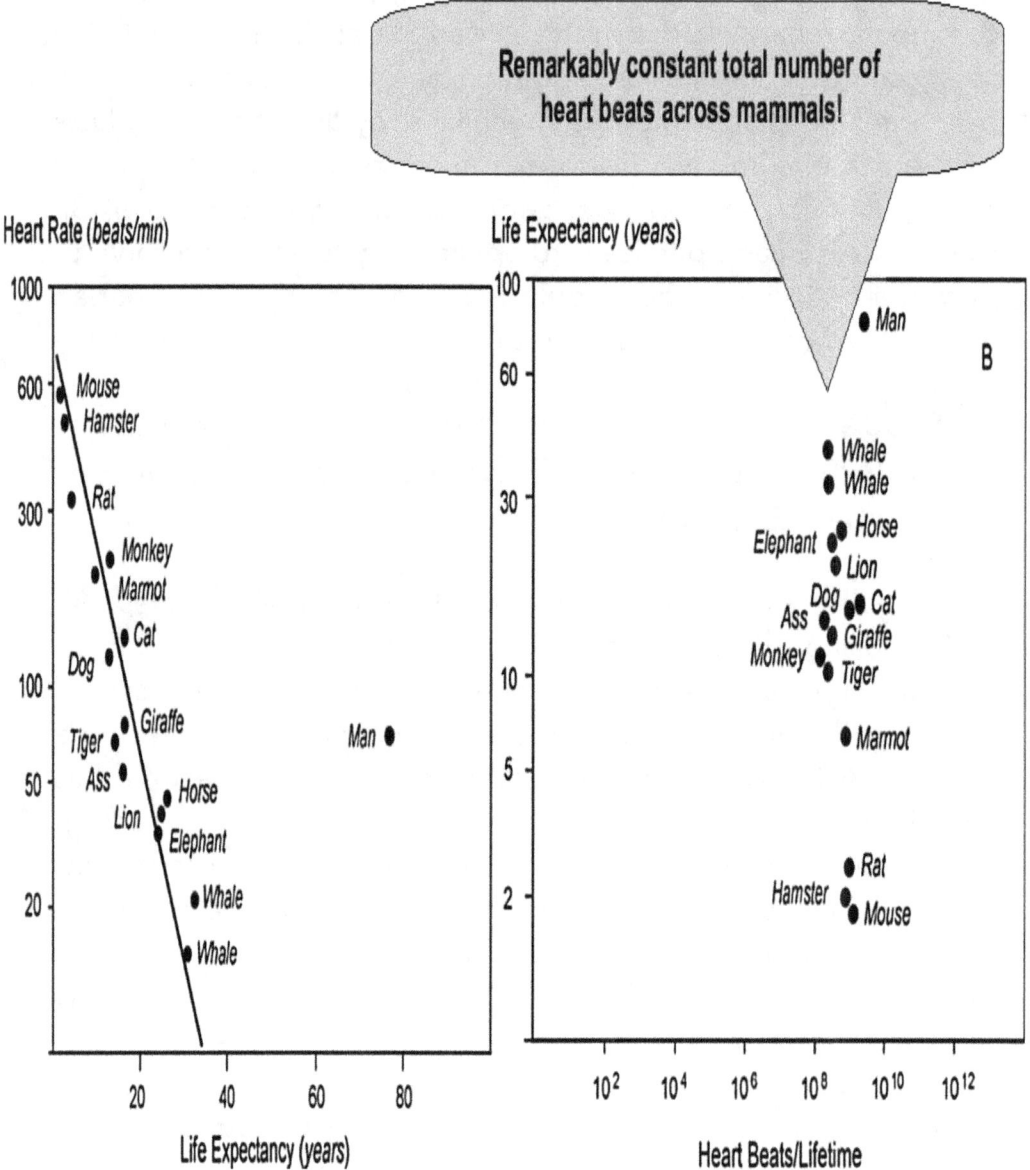

Figure A.2 Cardiac reserve limits longevity. Heart rate (left) as well as total number of heartbeats per lifetime (right) as well as heart rate correlate with life expectancy in mammals. (Figure modified from ref. 9).

Appendix: Applied Healthspan Engineering

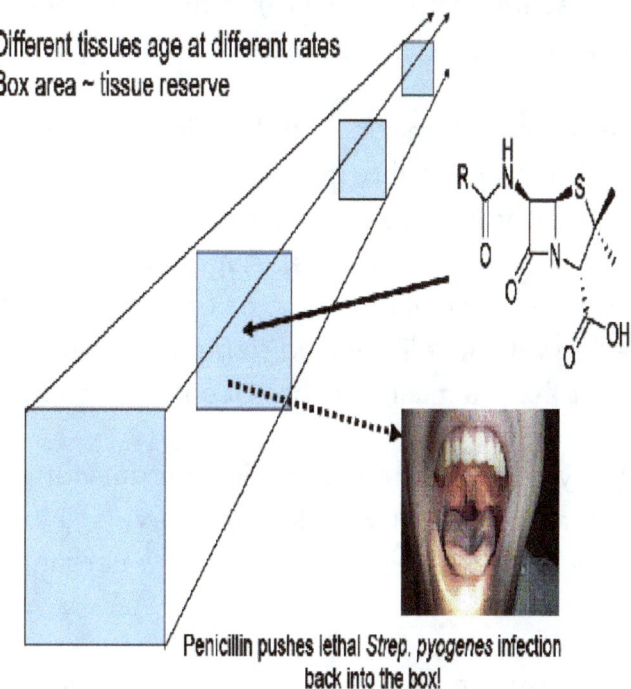

Figure A.3 Schematic depiction of decline in functional reserve with aging. Tissue reserve (illustrated by boxed area) decreases with age. Environmental stressors can push an individual (or an organ system) outside of the box. Therapy such as penicillin can restore well-being to a patient suffering from a life-threatening bacterial infection (inside the box).

conventional medical intervention is measured by its capacity to reduce the threat of injuries and conditions that push individuals "outside the box" of physiological reserve.

Solutions to the problem of inexorable loss of physiological reserve span a spectrum between two extreme cases. In the first case, an intervention slows decline of function in one or more organs. In the second case, function is restored or actually reversed. Restoration and reversal almost certainly involves replacement of cells, tissue, or organs. To date the replacement of hormones has provided short-term benefits with regard to well being with unclear and perhaps detrimental effects long term. Examples include growth hormone (15) estrogen replacement therapy, (16) dehydro-epiandrosterone (DHEA) (17-18) testosterone,(19-20), etc.

Inhibition of problematic cellular circuits (e.g., fibrosis, inflammation, etc.) is accessible today. In the future, it may be possible to stimulate protective cellular circuits or cell-based repair pathways (e.g., endogenous stem cells). Augmentation of cell-based repair pathways with new cells, tissues, or organs using stem cells and bioengineering will most likely lead to huge advances. However, limited progress at the practical level has been made to date.

Biological systems have a hierarchical modular structure that instructs interventions at any of several levels: Nutrients → macromolecular → organelle → cellular → tissue → organism (**Table**

TABLE A.1 Means of Intervention: An Artificial Hierarchy of Modularity

Category	Intervention
Spiritual?	Yoga, meditation, prayer, etc.
Physical	Temperature, electromagnetic, ultrasound
Nutrition/performance	Diet, calories, exercise,
Single molecules	Pharmaceuticals, nutraceuticals
Macromolecules	Proteins, fats, sugars
	Chaotic versus self-assembled ("nano-machines")?
Gene therapy	Therapy with "information"; genes, antisense, RNA interference
Organelle	Stimulation of autophagy, mitochondrial enhancement
Cellular	Therapy with "modular automata"; stem cells
Tissue/organ	Organ or tissue transplants, artificial organs/devices

A.1). On the basis of design and repair of various human-designed hierarchical systems (e.g., computers, automated manufacturing, machinery, etc.), the fastest and most facile way to restore high function is to replace the highest-level module above the malfunction. This approach works even when understanding is limited, as is the situation with aging. Thus, ultimately organ transplants and/or advanced regeneration methodologies may predominate AHE, as described in more detail below.

The physiological reserve of each organ or tissue declines at a characteristic rate. Reliable measurements of organ reserve and function serve as benchmarks to evaluate the interventions of AHE. Age-related decline in pulmonary function is illustrative (**Figure A.4, upper**) (21–24). This figure demonstrates age-specific decline in the forced expiratory volume in 1 s (FEV1), a standard measure of pulmonary function. Note that the rate of decline is accelerated by the habit of smoking cigarettes, a known insult to lung tissue. Although, cessation of smoking can reduce the loss of functional capacity with a major impact on public health, are there other interventions that can reduce the inexorable decline of pulmonary function and reserve?

Statins (drugs originally developed to inhibit 3-hydroxy-3-methyl-glutaryl-CoA reductase [HMG CoA reductase], the rate-limiting step in cholesterol biosynthesis) have been reported to slow age-related decline in pulmonary function (25,26). Results of the Boston Veterans Administration Normative Study (**Figure A.4, lower**), which followed a cohort of over 800 average age 71-year-old men for 10 years, revealed that use of statins significantly slowed the loss of pulmonary function (both FEV1 and lung capacity) in nonsmokers as well as previous and current smokers. At present, no interventions are known to reverse pulmonary decline except for lung transplantation, an intervention that is not practical on a large scale or at advanced age. Thus, statins are a rational component of AHE, acting to lower low-density lipoprotein cholesterol (LDL-C), to impede the progression of cardiovascular disease as well as multiple effects on inflammation, tissue repair, and homeostasis (27).

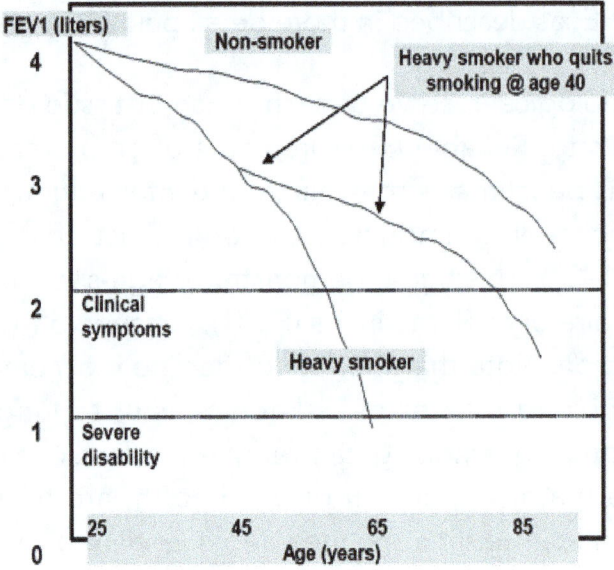

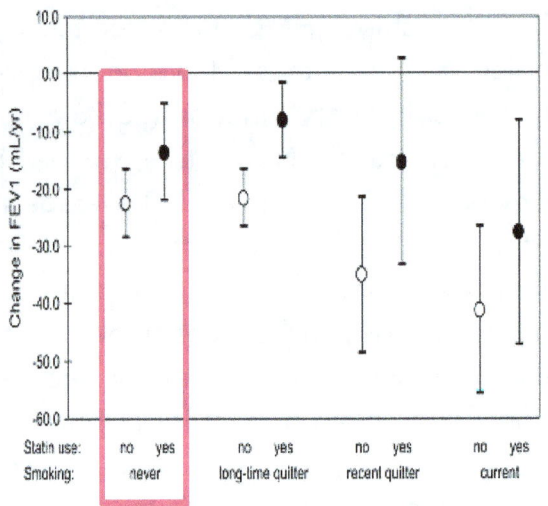

Figure A.4 Pulmonary reserve (forced expiratory volume, 1 s [FEV1]) declines with age (upper); however, decline can be slowed by statin therapy (lower). (Top) FEV, a measure of pulmonary function, decreases with age. (Based on data from ref. 20). (Bottom) Statin use slows decline in FEV in 803 men with an average age of 71. Forced vital capacity (FVC) and FEV1 were measured two to four times between 1995 and 2005. (Figure modified from ref. 25).

What Interventions If Any Can Slow End Organ Deterioration?

Biological systems are comprised of several levels or hierarchies of modularity, e.g., molecular, organelle, cellular, tissue, and organism (**Table A.1**). Although functional "wellness" can be measured at the organismal and tissue levels, the underlying biochemical processes act across modules, complicating interventions. Although interventions to slow or reverse dysfunction can be made at any of several levels, there is an imperfect match between type of intervention and "module/level of action" (see **Table A.1**). As noted above, two options are available:

(1) modify native processes or

(2) simply replace a particular module.

Extensive data support the benefits of diet and exercise to increase healthspan. Such interventions are easily and immediately accessible. Unfortunately, motivation may limit efficacy, hence the interest in an exercise pill (28).

Prophylactic alteration of the activity of disease-associated physiological pathways is not without the possibility of tradeoffs in which suppression of inflammation, for example, may reduce wound healing and immunity. Even life-saving therapeutics may actually accelerate aging. For example, cancer chemotherapy may induce additional tumor mutations, increasing drug resistance, and contribute to development of subsequent secondary neoplasms. Chemotherapy may also cause the destruction of adult stem cell populations that help maintain the organism and delay aging.

Elimination of exposure to stress and exogenous infection/inflammatory agents/stimuli is moderately controllable, especially by vaccination in some cases. However, substantial progress will be required to eliminate chronic viral infections that apparently contribute to long-term debilitation of the immune system. The continual evolution of pathogens and difficult development of new, engineered substances make this approach problematic.

Certain surgical procedures, although invasive, effectively eliminate, delay, or repair certain conditions, such as degenerative diseases of the heart or cardiovascular system. Improved surgical approaches employing micro- and nano-devices are under development and may play an important role in the not-so-near future.

Modulation of Physiological Processes to Extend Healthspan (Targeting Specific Critical Tissue/Organ Systems

Nutritional supplementation and/or pharmacological interventions may partially overcome genetic limitations of metabolism by attenuating or stimulating key pathways. Presently, despite substantial supplement industry advertising, there are no commercially available nutraceutical or pharmaceutical interventions proven to slow the aging process substantially in humans. However, a strong rationale may exist for some agents. Furthermore, in the near future, personalized medicine based on genome analysis may permit development of tailored prophylactic antiaging, pro-healthspan regimens (**Table A.2**) (29–32).

TABLE A.2 Representative Rational Healthspan Interventions

Target/process	Intervention	References
Blood pressure	Multiple; exercise, dietary, sodium restriction, see RAS (below)	48
Heart rate	Exercise, vagal nerve stimulation	9
Dyslipidemia	Fish oil; flaxseed oil, olive oil niacin, statins	34–37, 48–51, 123–125
Renin–angiotensin system (RAS)	Exercise, dietary, sodium restriction, ACE inhibitors, ARBs, renin inhibitors	See Table 3
Medial elastocalcinosis	Vitamin K2	33
Glucose homeostasis	Exercise, metformin, dietary-caloric restriction	42, 47, 52, 53
mTOR pathway	Resveratrol, rapamycin, dietary-caloric restriction	41, 42, 43–46, 54
Inflammation	Aspirin, NF-κB inhibitors (e.g., EGCG, quercetin, etc.)	55, 58
Autophagy	Verapamil, trephalose, others	56, 57
Extracellular matrix cross-link	Alagebrium, ALT-711	59
Chemopreventive	Aspirin, bioflavonoids	38–40, 60

ACE, angiotensin converting enzyme; ARBs, angiotensin receptor blockers; EGCG, epigallocatechin 3-gallate; mTOR, mammalian target of rapamycin.

Increased understanding of the pathophysiology of mammalian/human diseases associated with, but probably secondary consequences of, fundamental aging processes provide a very good set of targets to extend healthspan. Although many components of these pathways are evolutionarily conserved and may be studied in invertebrates, the significant differences in metabolism, as well as their effects on the homeostasis of complex mammalian organ systems, will require development of mammalian/human model systems to advance AHE. Several pathways accessible today are described below.

The renin–angiotensin system (RAS) is among the most important physiological mechanisms maintaining homeostasis of the cardiopulmonary system. Although of paramount importance in regulation of blood pressure, the RAS has global effects, helping to regulate tissue maintenance in almost every organ (**Table A.3; Figure A.5**) (85)

Example #1: Renin–angiotensin system

Readily available drugs (e.g., angiotensin converting enzyme [ACE] inhibitors and angiotensin receptor blockers [ARBs], that inhibit this pathway have been shown to extend rodent lifespan (86). Recent work of Basso *et al.* (86) demonstrated a ~20% increase in median and maximal lifespan in rats treated with enapril (ACE inhibitor) or losartan (ARB). Although these investigators documented a significant reduction in systolic blood pressure (&144 mm Hg in controls vs. ~105 mm Hg in treated), most impressive was the reduction in vascular fibrosis. Histology figures in their report (http://ajpheart.physiology.org/cgi/content-nw/full/293/3/H1351/F1-F3) show dramatic reductions in tissue fibrosis in the treated animals compared to controls. Presumably this reduction in tissue fibrosis contributes to increased physiological reserve and function in multiple tissues. Consistent with these findings, mice lacking the AT1A receptor, the target of ARBs, live longer than littermate controls (87). Given the significance of cancer in mouse longevity, the primary mechanism of action may be on reduced tumor angiogenesis and growth (88, 89). Alternatively, kidney function is also responsible for age-associated mouse (90) and rat deaths (91), and lower blood pressure associated with inhibiting the RAS may also reduce kidney damage. Recent data suggest that cardiac function, which is also adversely affected by angiotensin signaling (73) may be partially responsible for mouse deaths (89, 92). AT1 angiotensin receptor inhibitors have been reported to inhibit the formation of circulating MDA-modified LDLs in human diabetic patients and the modification by MDA of lungs in rats affected with bleomycin-induced pulmonary fibrosis (93). **Table A.3** summarizes the benefits of RAS blockage that, in addition to those already described, include anti-atherosclerotic,

TABLE A.3 RENIN–ANGIOTENSIN SYSTEM BLOCKADE: PROTECTIVE ACTIONS IN MULTIPLE ORGAN SYSTEMS

Pathology	Treatment	References
Hypertension		
Vascular dysfunction	ACEi, ARB	61, 81
Atherosclerosis		
Excess adhesion molecules	ACEi	62
Arterial stiffness	ACEi, ARB	63–66
Arterial fibrosis	ACEi, ARB	67, 68
Arterial hypertrophy	ACEi, ARB	67, 68
Endothelial function	ACEi	69, 70
LDL oxidation	ACEi, ARB	71, 72
Heart		
Myocardial fibrosis	ACEi	73
Ventricular hypertrophy	ACEi	74, 75
Atrial fibrillation	ACEi, ARB	76–78
Diabetes		
Insulin resistance (type II diabetes)	ACEi, ARB	79, 80
Diabetic nephropathy	ACEi, ARB	81, 82
Inflammation		
Inflammatory cytokines	ACEi	69, 70
Low levels bradykinin, NO	ACEi	69
Monocyte activation	ARB	83
Autoimmune		
Autoimmune encephalomyelitis/ Multiple Sclerosis	ACEi	84

ACEi, Angiotensin-converting enzyme inhibitors; ARB, angiotensin II receptor blockers; LDL, low-density lipoprotein; NO, nitric oxide.

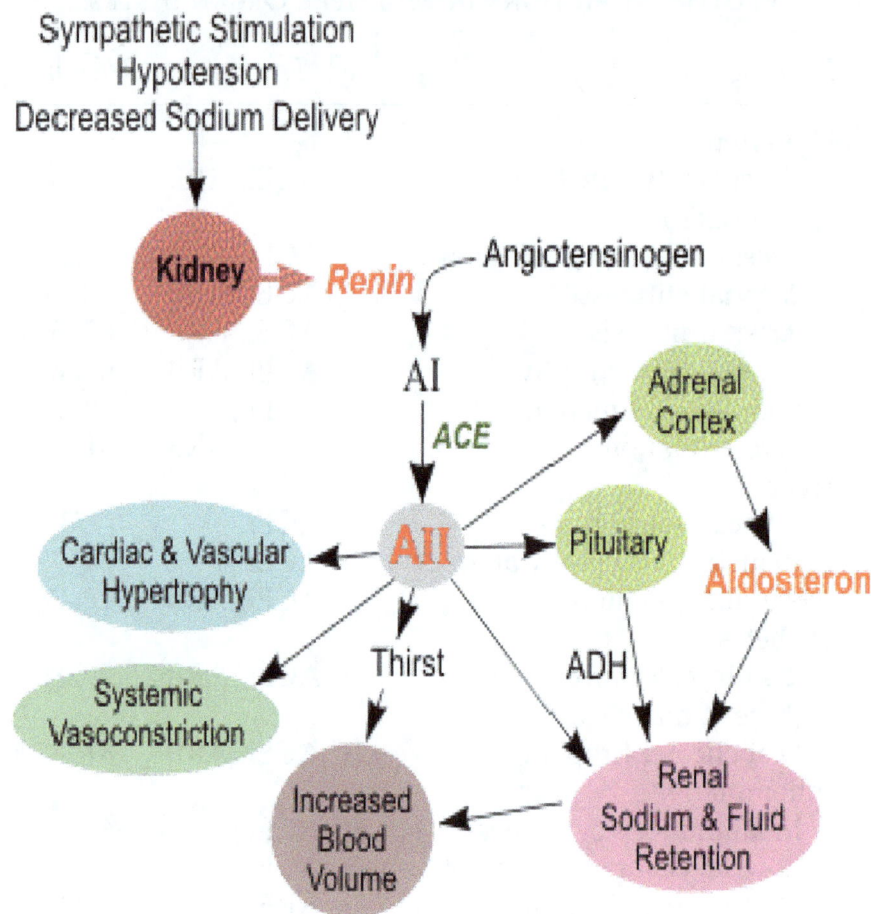

Figure A.5 Renin–angiotensin system physiology. AI/AII, Angiotensin I and II; ADH, antidiuretic hormone; ACE, angiotensin-converting enzyme.

anti-inflammatory, anti-autoimmune, and procardiac activities.

In summary, the RAS, so critical for mammalian homeostasis, cannot be studied in models of aging in yeast or invertebrates, yet is quite important to mammalian healthspan and lifespan.

Example #2: Medial elastocalcinosis

Medial elastocalcinosis is a significant process within vascular tissue that contributes to cardiovascular decline with aging (94–96). This process is modulated by several vitamin K–dependent, gamma-carboxylated (Gla) proteins, e.g., matrix Gla protein (MGP), osteocalcin, growth arrest specific gene 6 (Gas6), and four transmembrane Gla proteins (TMGPs). Recent work of Schurgers *et al.* demonstrated a significant reduction in medial elastocalcinosis of rodents given pharmacological doses of vitamin K2 (menaquinone) (33). Figures in their original report (bloodjournal.hematologylibrary.org/cgi/content/full/109/7/2823/F2-F6) show reversal and up to 50% reduction in vessel calcification. Supplementation with vitamin K2 is a relatively benign means to reduce vascular calcification.

Example #3: Lipid metabolism

Lipid metabolism plays a critical role in organismal homeostasis, particularly the cardiovascular system. Statins are the frontline drugs to reduce LDL-C to slow or reverse atherosclerosis and maintain cardiovascular health (34).Many beneficial effects result from the pleiotropic effects of statins: Protection of endothelial function via augmentation of nitric oxide, increasing the number of endothelial progenitor cells, decreasing oxidative stress via reduction of nicotinamide adenine dinucleotide phosphate (NAPDH) oxidase, reduced thrombogenic response via decreased platelet function, decreased lymphocyte activation/adhesion and decreased inflammation as measured by reduced C-reactive protein (see Figure 1 in ref. 45)(36). Statins also have modest beneficial effects on atrial fibrillation, the most common arrhythmia of the elderly (97). Benefits for heart failure (98,99) and complications of type II diabetes (100) are under investigation. Substantial evidence supports using statins to prevent cardiovascular disease and augment cardiovascular reserve (101). Fish oil containing docosahexaenoic acid (DHA) is now well known to benefit cardiovascular health by reducing blood pressure, heart rate (37) and function of platelets (102). Other benefits include attenuation of the activity of immune-mediated inflammatory pathways that contribute to increased incidence of age-associated diseases from arthritis, atherosclerosis, and cancer to Alzheimer's disease.

Example #4: Anti-inflammation

Chronic systemic inflammation increases after middle age in humans, resulting in tissue damage that reduces reserve and favors tumor formation (103,104). Many diseases associated with aging, such as Alzheimer's disease (105), rheumatoid arthritis (106), chronic obstructive pulmonary disease (107) and atherosclerosis (108) have significant immunoinflammatory components. Increased serum levels of inflammatory cytokines tumor necrosis factor-alpha (TNF-a, interleukin-1 beta (IL-1b, as well as the associated anti-inflammatory cytokine IL-6, are often found elevated in older people (103). Body mass index increases with age (103) and increased adipose tissue may explain the increased cytokine production (109). Therefore, a balanced diet and exercise to control weight can provide significant benefit to control inflammation. Many of the secondary benefits of statins, fish oil, ARBs, and ACE (Examples 2 and 3) are hypothesized to be due to suppression of inflammation (110,111). Numerous studies support the use of low-dose acetyl salicylic acid (ASA, aspirin) to attenuate augmented platelet function that contributes to vascular disease (112). ASA at higher doses may be chemo-preventive for colonic tumors (38–40). Nonsteroidal anti-inflammatory drugs (NSAIDs) such as ibuprofen have been reported to decrease the incidence of Alzheimer's disease in some studies, although further study is needed (113,114). Reactive oxygen species (ROS) play an important role in the induction and maintenance of systemic inflammation, suggesting that antioxidant therapy may be of use as well (see below). Preventing or ameliorating potential tissue damage caused by activation of inflammatory pathways is an important goal of AHE. The reduction of body mass index may play at least as important role as pharmaceutical or nutraceutical intervention.

Other Metabolic and Physiological Pathways Suitable for AHE

For ultimate success, AHE requires an inventory of the pathways involved in organ maintenance and decline. What processes limit physiological reserve? A deep understanding of the limits of cellular and tissue homeostatic mechanisms is necessary. Unfortunately, for integrated organ–tissue systems, maintenance requires continuous high-level function from specialized, often postmitotic cells. A tension will arise pitting the limits of pathway augmentation against the need to replace aged, malfunctioning cells *in toto*, i.e., bona fide regenerative medicine.

Growth/stress pathways: Universal maintenance/ longevity pathways?

Alterations of specific pathways that control growth, homeostasis, and stress response have been shown to increase longevity in yeast, *Caenorhabditis elegans*, *Drosophila melanogaster*, and mice. Inhibition of the evolutionarily conserved insulin/insulin-like growth factor-1 (IGF-1)/phosphoinositide-3 (PI3) kinase/Akt pathway or TOR pathway results in increased longevity, presumably through an incompletely understood combination of growth suppression by decreased ribosomal biogenesis and protein translation and increased repair by autophagy (41). Dietary or caloric restriction (CR), which appears to act through these two pathways, increases lifespan and healthspan in all of these organisms. It has been hypothesized that CR will similarly increase healthspan and lifespan in humans and other long-lived primates. Although data do not yet support lifespan extension, CR has significant healthspan benefit in primates and humans, including insulin sensitization and improved cardiopulmonary function (42,115). The benefits of CR points to the existence of cellular maintenance mechanisms that may not only be ubiquitous in eukaryotes, but likely can affect all organ systems to some extent. However, the existence of universal mechanisms of cellular maintenance contributing to longevity does not obviate the need to discover and reverse defects in homeostasis associated with specialized tissues, e.g., kidney, eye, brain, heart, etc.

Recent interest in the development of CR mimetics to increase healthspan/lifespan has skyrocketed. Headlines have been captured by two substances. The mTor inhibitor rapamycin increased lifespan in mice, although at the possible cost of increased rates of lymphoma (43). Resveratrol, a bioflavonoid available as a dietary supplement, activates Sirt1, a downstream effector of the CR pathways. Resveratrol has been reported to protect mice from damage due to consumption of high fat diets (44,45). Many of the same patterns of gene expression seen with CR are found after exposure to resveratrol (46).

What about antioxidants? (see also Chapter 10)

Accumulation of damaged macromolecules is hypothesized to underlie loss of cell function with aging. Such damage has been linked to highly reactive chemical species, such as free radicals of oxygen (ROS) or nitrogen. Harman first postulated, the "Free Radical" and later the "Oxidative Stress" theory of aging in 1956 (116). These concepts led to the conjecture that substances with antioxidant activities, such as free radical scavengers, would limit the generation and damage mediated by free radicals. These interventions were expected to help maintain normal cellular function and thereby slow aging. This catechism has become a dogma among the general population with >40% consuming antioxidant supplements of one type or another (117). A huge industry has grown up around this idea.

However, there is limited evidence that antioxidant supplementation, outside of preventing deficiencies, provides any positive effect on longevity or the maintenance of healthspan in healthy individuals (118,119) In fact, it appears that in some cases antioxidants may actually inhibit normal ROS and damage protective mechanisms by eliminating the oxidative stress that would normally trigger their induction. For example, Ristow et al. (120) showed that supplementation with antioxidant vitamins C and E blocked the beneficial increase in insulin sensitivity that is associated with moderate exercise, a known enhancer of healthspan (47).

Many other reports question the validity of the free radical theory of aging (reviewed in ref. 122). With the exceptions of *Drosophila* (121) and mitochondrial localization of catalase in mice (89,92,123), neither antioxidant vitamin supplementation nor individual ectopic expression of protective enzymes in their endogenous subcellular locations (e.g., catalase and superoxide dismutase) have been shown to increase longevity in rodents and *C. elegans*, although these organisms may be better protected from exogenous oxidative stress, e.g., paraquat (122). Remarkably even combined endogenous expression of a superoxide dismutase and catalase does not extend longevity in mice (124). Despite exhibiting increased oxidative stress, *Mclk*$^{+/-}$ mice that are partially coenzyme Q (CoQ7) deficient live longer than normal littermate controls (125). Increased oxidative damage is also observed in the long-lived naked mole rat, compared with short-lived rodents such as mice or rats (126). Consistent with these studies, Van Raamsdonk et al. demonstrated that the sod-2 mutant of *C. elegans*, has increased oxidative stress, but lives longer (127).

There could be many reasons for the failure of antioxidants to produce a significant increase of healthspan or lifespan. The simplest explanation may be that the most studied antioxidants either lack enough broad-spectrum activity or exhibit counterproductive activities that interfere with potential benefits. For example, many antioxidants, such as ascorbic acid, are redox compounds that when oxidized can act as prooxidants. For example, transport of increased antioxidant vitamin C to the murine lens actually causes increased protein cross-linking and cataract formation (128).

Alternatively, failure of antioxidants to provide significant benefit may be much more fundamental. ROS are not only a potential source of damage, but are also employed by evolutionary processes in eukaryotic development and homeostasis. ROS signaling through conserved NADPH oxidases promotes cell proliferation, maintenance of the undifferentiated state in pluripotent stem cells, new blood vessel formation, and developmental processes such as cardiogenesis. Furthermore, the innate immune system uses ROS generation by NADPH oxidases to destroy pathogens (129). Most significantly, activation of NADPH oxidases plays a significant evolutionarily conserved role in wound repair (129). Inappropriate activation of wound-healing pathways by age-associated disease processes activate angiotensin II and proinflammatory cytokines that in turn stimulates NADPH oxidase activity. Resulting inflammation can exacerbate incipient diseases associated with aging such as arthritis, fibrosis, cancer, and reperfusion injury due to stroke. Although antioxidants may not provide systemic relief of age-associated decline in organ reserves, they have proven useful in attenuating inflammatory signaling associated with NADPH oxidase activation in several diseases. Two antioxidants are used clinically. N-Acetyl cysteine is used to treat acetaminophen toxicity. The free radical scavenger edaravone is used clinically to reduce damage caused by activation of the arachidonic acid cascade associated with reperfusion injury of the brain (130). Because of the wide spectrum of inflammatory diseases, antioxidants may still have an important role to play in medicine and AHE.

Activation of inflammatory pathways by NADPH oxidase signaling provides a relatively simple framework to reinterpret experiments that seem to support the free radical theory of aging. For example, we hypothesize that mitochondrial-directed catalase expression in transgenic mice (123) suppresses inappropriate activation of the RAS/NADPH oxidase/fibrotic axis. Reduced RAS activity may make a greater contribution to the increased longevity of these mice than the observed decreased oxidative damage, (89,92,123) although it is difficult to distinguish these possibilities given that damage can lead to a ROS-based positive feedback signaling loop. In this explanation, increased angiotensin II in the cardiovascular system of aging mammals (131) stimulates fibrosis/hypertrophy by activating cytoplasmic NADPH oxidase (132), which in turn stimulates mitochondrial ROS by opening the mitochondrial permeability transition pore. Mitochondrial ROS then activates p38/ERK1/2/MEK1/2 kinases (133) and profibrotic pathways resulting in cardiac dysfunction (134). Although the resulting oxidative damage is secondary to RAS signaling, this damage probably plays a role in the reduced cardiac function, possibly by inducing DNA oxidation and subsequent stress-mediated cell senescence (135). Consistent with this interpretation is that knockout of angiotensin II receptor AT1A not only increases murine longevity but also decreases oxidative markers in many organs, including the heart (87).

Another way to reconcile the relatively poor efficacy of antioxidants is to invoke a more generalized version of the oxidative stress model of aging. In this model, generalized accumulation of "molecular garbage" is postulated to be responsible for reduced cell function that may eventually result in a "garbage catastrophe."(136). This hypothesis incorporates more potential sources of damage than the simple ROS theory. For example, unfolded proteins and inappropriate macromolecular modification (e.g., advanced glycation end products) are included (where ROS may not be the root cause). Substances that promote reduced garbage accumulation or better garbage disposal would be the target of new small molecular drug/nutraceutical development. A key function may be to stimulate autophagy or induce a stronger unfolded protein response in addition to any antioxidant function.

The loss of homeostasis with resultant compromise of physiological reserve after the end of the reproductive period is more than simple garbage accumulation. We hypothesize that there are limitations of design inherent in somatic homeostasis. Two examples are dysregulated gene expression with time ("developmental drift") or absence of sufficient rejuvenative and/or regenerative mechanisms. An example of the former is the increased activity of GATA transcription factors elt5 and elt6 in aged *C. elegans*, leading to decreased expression of the master regulator elt3. Inhibiting elt5 and elt6 activity restores elt3 function and extends worm lifespan by up to 50% (137). Adult *C. elegans* are postmitotic, with limited ability to replace damaged or aging cells (138). This compromised regenerative capacity may be fundamental to the observed reduction in physiological reserve and proximate demise of elderly worms. Our hypothesis not only explains the inability of antioxidants to significantly enhance lifespan/healthspan, but also suggests the importance of such design limits on longevity itself. The conclusion may mandate cell-tissue-organ replacement strategies as discussed below.

Future Directions: Enhanced Regeneration and Replacement Strategies for AHE

Natural selection also makes extensive use of replacement in organ maintenance in long-lived organisms such as humans. Replacement involves proliferation of progenitor or stem cells. In humans, there is a great amount of tissue-specific cell turnover. For example, new skin cells, gastrointestinal epithelial cells, and hematopoietic cells are produced continually. At the other extreme, new skeletal muscle, cardiomyocytes, and neurons are rarely formed, although in each case a limited capacity for renewal results from embedded stem cells in adult tissue. Although both progenitor and stem cells may be involved in organ homeostasis, a distinction can be made in tissues that have high cell turnover rates, such as tissue composed of epithelial cells. In the lung, progenitor cells maintain homeostasis, with an overall proliferation rate of less than 1% per day (139, 140). Clara cell secretory protein expressing stem cells are responsible for resistance to pollutants and regenerate specialized epithelial cell types in the bronchioles. Such cells function to repair more serious injuries that result in local depletion of epithelial progenitor cells (141). Moreover, some tissues, such as pancreatic islet cells, are believed to lack stem cells and solely use progenitor cells for maintenance (142).

Conversely, natural selection of organisms that lack replacement mechanisms are largely postmitotic as adults and are short-lived, as is the case of C. elegans (138). Because organs are not needed after the period of successful reproduction, mechanisms to maintain organ function past this time are not likely to have been selected. This likely explains the small number of cardiac stem cells in the adult mammalian heart and the low level of cardiomyocyte turnover during the human lifespan (143).

Enhanced regeneration appears to be the mechanism used by natural selection to maintain viability in the few examples in the animal kingdom in which the soma is not disposable. Asexual species of planarians and hydra lack germ cells, but instead have pluripotent somatic cells to replace aging, postmitotic tissue. Hydra eliminate aging cells by sloughing off differentiated cells and are considered essentially immortal (144). Asexual planaria detach their tails and regenerate two halves of their body, a process that requires their pluripotent somatic stem cells, neoblasts, to not only differentiate, but to recreate the positional information inherent in their body plan (145). Although the planarian's method of maintaining healthspan is not practical for humans, the idea that it may be possible to significantly enhance tissue regeneration has significant merit.

AHE seeks replacement strategies that include enhanced rejuvenation/regeneration, stem cell–based tissue/organ repair, tissue/organ replacement, and the development of durable artificial organs.

Enhanced regeneration

Stimulation of endogenous tissue maintenance pathways to remove poorly functioning cells and replace them with new cells derived from progenitor or stem cells is the least disruptive strategy for tissue rejuvenation. Successful development of this strategy necessitates avoidance of dysfunction cellular immortalization (cancer) and replicative senescence (Hayflick limit).

Enhanced regeneration is likely to be highly tissue specific. Blastema formation, a hallmark of limb regeneration in newts and axolotls, has been observed in punctured ears of MRL mice (146) as well as other strains (147). Thus, mammals may possess far more extensive regenerative capacity than is usually believed, although it appears that critical organs such as the heart and brain possess very limited regenerative capacity.

Until recently, the idea that the heart had significant regenerative potential or the capacity to generate new cardiomyocytes was controversial. Recent evidence suggests the human heart does have limited regenerative capacity that might be exploited for AHE. Hsieh *et al.* have provided evidence for the existence of at least a small population of adult cardiomyocyte stem cells that replace damaged cardiomyocytes after injury (148).

Furthermore, human cardiomyocytes are not a static postmitotic population. Bergmann *et al.* used carbon-14 (originating from 1950s atmospheric nuclear bomb tests) levels in DNA of human cardiomyocytes to demonstrate cell proliferation ranging from 1% in 25 year olds to 0.45% in 75 year olds (143). This implies that almost 50% of cardiomyocytes may be renewed during a human lifetime.

The first promising step in the development of pharmaceutical rejuvenation techniques to increase new cardiomyocyte formation has been recently described. Stimulation of mononucleated cardiomyocytes (most cardiomyocytes are binucleated) by neuregulin-1 (149) causes up to 6.5% of mononucleated cells to undergo DNA synthesis. Sufficient cardiomyocyte replacement prevented cardiac dilatation in a murine congestive heart failure model (following ischemia reperfusion induced by ligated left anterior descending artery [LAD]). Cell cycle reentry of postmitotic cells may be possible in skeletal muscle as well. A small synthetic molecule, reversine, stimulates skeletal muscle to dedifferentiate into a mesenchymal-like stem cell capable of forming chondrocytes and adipocytes (150).

Development of new drugs that enhance preexisting regenerative capacity are an important goal of AHE, although serious hurdles persist. At a minimum, tissue specific/ localized delivery may be required to contain the potentially deleterious activities of such compounds.

Stem cells and artificial organs

Stem cells provide a potential means to replace or augment old or malfunctioning tissue. In principle, physiological reserve and maintenance of homeostasis could result. However, four basic problems need to be solved for optimal use of stem cells:

(1) means to introduce the cells into a target tissue,

(2) means to insure correct cell differentiation,

(3) means to re- move damaged or dysfunctional cells, and

(4) means to insure integration into the target tissue.

Replacement of damaged or diseased tissue is already being attempted with a variety of adult stem cells. Sources include hematopoietic, bone-marrow derived mesenchymal cells, and adipose cells.

Given the heart's limited regenerative capacity, a variety of attempts have been made to treat heart failure resulting from myocardial infarction using injection of either hematopoietic or bone marrow-derived mesenchymal autologous stem cells. Improvement of up to 10% of left ventricular ejection fraction has been observed. It appears that these adult stem cells mainly increase blood supply to the heart by differentiating into endothelial cells. The problem of delivery and retention of stem calls in the area of injury can be addressed by bispecific antibodies, one binding to the stem cell and the other binding an injury antigen such as myosin light chain (151). Unfortunately, this approach does not solve the cell differentiation problem, i.e., provide viable replacement cardiomyocytes.

The recent development of techniques to reprogram somatic cells to become pluripotent (induced pluripotent stem [iPS] cells) may be the breakthrough needed to generate sufficient numbers of specialized autologous cells for repair and replacement of damaged tissue. Takahashi (152, 153) and colleagues found that ectopic expression of four transcription factors, Oct4, Sox2, Klf-4, and c-Myc, could reprogram somatic mouse and human tissue into cells that resemble embryonic stem (ES) cells. The original methodology used retroviral and lentiviral gene therapy techniques that are currently problematic for human therapy. Presently there is no gene therapy treatment approved as a therapeutic in the United States. However, recent work suggests a combination of small molecule drugs and cell-transducible reprogramming proteins will increase the safety of iPS production (154–156). Techniques to differentiate iPS cells into many cell types, including cardiomyocytes, are under active development (157) and clinical trials are on the immediate horizon.

Despite substantial progress, outstanding problems remain. Among these are: How best to remove dead or damaged cells without scar formation, how best to remove aged cells, and how best to promote cellular integration without disrupting tissue function. The latter problems are likely to be more difficult than the former, because the human body has mechanisms for removal of dead cells that can be co-opted.

Haphazard cellular integration may be problematic because appropriate structural and electrical connections must be established in the heart. Unfortunately delivery of skeletal muscle cells, originally thought to be a ready source of large numbers of autologous repair cells, failed to revive damaged hearts due to associated serious arrhythmias (158). Transplanted cells can apparently sense potential problems with tissue integration; as a result, cardiomyocytes directly injected into rodent hearts have a strong tendency to die before successful integration (159).

iPS cells and ES cells are also being used to create whole organs ex vivo for transplantation. Although this goal remains in the realm of science fiction, various hybrid organs are being developed using a combination of differentiated cells derived from ES or iPS cells, bioengineered microenvironments, scaffolds, and encapsulation materials. For example, substantial progress has been made toward an artificial liver (160).

A serious drawback for AHE is that organ replacement

requires transplantation surgery. Although iPS cells solve the immune rejection problem associated with current techniques, incomplete reenervation and the risks of surgery, especially on the elderly, make such a strategy impractical. Delivery of new tissue by catheterization may reduce this risk, although organ size limits this approach. Perhaps cells and tissues can be programmed for self-assembly in vivo. For example, heart valve engineers are designing possible in vivo assembly strategies in conjunction with catheter-based delivery (161).

Bioengineered mechanical organs

Although AHE seeks to optimize biologically based tissues and organs ("carbon-organic-based solutions"), it is necessary to mention that competing biodevice technology ("silicon-inorganic based solutions") may ultimately prove superior. For example, the implementation of an implantable artificial heart has been a holy grail of bioengineering for quite some time (162). Successful development of left ventricle assist devices (LVAD) represents an important milestone with survival of some patients for months to years. Current problems include strokes, unwieldy power connections, anticoagulation, and increased risk of potentially fatal infections. Interestingly, implantation of LVADs actually restored native heart function in some cases where a transplantable heart was not identified (reviewed by Mountis and Starling, 163).

Problems of integration of artificial biomaterials with biological systems are likely to be solved by hybrid approaches. We hypothesized that repair is best implemented by replacing the highest-level module above the identified malfunction (**Table A.1**). This approach requires the least understanding of the underlying system components. Thus, stem cell–based and biomechanical organ transplantation may be predicted to advance more rapidly than technologies impinging at lower levels (e.g., pharmaceuticals).

Potential problems with regenerating the brain

The brain poses a special problem for AHE replacement strategies. Whole brain replacement is obviously nonsensical (although head transplants have been patented!; e.g., US Patent 4,666,425), and even replacement of individual neurons by regenerative therapies may be problematic for maintenance of memory and identity, depending upon the role some individual neurons play in distributed memory networks (164). The creation and maintenance of memory remains a topic of active research, and the degree to which loss of specific neurons or addition of new neurons will disrupt brain memory can only be crudely estimated from brain injury patients.

Although the brain possesses limited homeostatic regenerative capacity, neurogenesis occurs in the hippocampal dentate gyrus and the subventricular zone of the lateral ventricles. The function of this regenerative capacity in adults is not completely understood, but may play a role in spatial memory formation (reviewed in Lee and Son,165). Neurogenesis can be stimulated by a variety of compounds in mammals, including dietary supplements such as curcumin (166). However, it is especially interesting that conventional aerobic exercise is among the most effective means to stimulate neurogenesis (167). Although it is unclear to what extent these cells contribute to maintenance of brain homeostasis, exercise may provide a low-tech means to achieve increased maintenance of brain function. Certainly the cardiopulmonary benefits of aerobic exercise on brain function cannot be ignored.

Conclusion: "Meet Me at the Salad Bar after We Go for a Jog"

Unfortunately, a large gap separates potentially useful AHE interventions and bona fide evidence of significant healthspan benefit. Improved perinatal health, sanitation, and elimination of many serious life-shortening infectious diseases were most responsible for the increased healthspan of the 20th century. Ironically, despite huge advances in our understanding of fundamental metabolic processes and billions of dollars spent on discovery and development of new drugs, only modest pharmaceutical recommendations can be made presently. Consuming a modest balanced diet consumed in modest meals (forestalling type II diabetes and atherosclerosis) and partaking in daily exercise (causing new blood vessel formation, neurogenesis, reduced heart rate by vagal nerve tone, reduced inflammation, etc.) presently have the greatest potential to maintain critical organ reserves. Unfortunately, worldwide "overnutrition" coupled with modern sedentary habits are a significant roadblock to successful AHE. The present generation may be the first in almost 200 years to have a shorter median lifespan than its parents!

Fundamental questions remain regarding tissue homeostasis and maintenance of organ reserve. What is the universe of pathways that control homeostasis? What is the hierarchy of pathways? Which pathways are most critical? What are the limits of intervention? Development of novel biomarkers of aging are needed to guide future studies. However, inherent limitations of mammalian cellular and tissue design create barriers to highly successful AHE. Molecular interventions will likely provide only partial, incomplete solutions for AHE, hence the need for regenerative medicine.

Regenerative medicine promises to revolutionize AHE. We advocate an AHE-oriented research approach focused on the mechanisms of organ/tissue maintenance with age. Unfortunately, numerous pathways important in human healthspan are not well modeled in short-lived species used to study longevity. New model systems will need to be developed to measure functional decline in the absence of disease for critical systems such as the heart and lungs. Perhaps the decline in cardiac function seen in old C57Bl6 mice will make a good starting point to model the heart (92). These model systems could be used to elucidate the limits of tissue maintenance by progenitor and stem cells. They would be useful to discover new targets for intervention and the limits to in situ stimulation of regeneration versus wholesale cellular or organ replacement. By whatever means, we anticipate that successful AHE will square the healthspan curve and greatly enlarge the population of youthful centenarians.

References

1. Homeric Hymn to Aphrodite (trans. H.G. Evelyn-White). Hesiod: The Homeric hymns and homerica. Harvard University Press, Loeb Classical Library, Cambridge, MA; 1981.
2. Aguilaniu H, Gustafsson L, Rigoulet M, Nystro m T. Asymmetric inheritance of oxidatively damaged proteins during cytokinesis. Science 2003;299:1751–1753
3. Murga M, Bunting S, Montana MF, Soria R, Mulero F, Canamero M, Lee Y, McKinnon PJ, Nussenzweig A, Fernandez-Capetillo O. A mouse model of ATR-Seckel shows embryonic replicative stress and accelerated aging. Nat Genet 2009;41:891–898.
4. Tang WY, Newbold R, Mardilovich K, Jefferson W, Cheng 477. RY, Medvedovic M, Ho SM. Persistent hypomethylation in the promoter of nucleosomal binding protein 1 (Nsbp1) correlates with overexpression of Nsbp1 in mouse uteri neonatally exposed to diethylstilbestrol or genistein. Endocrinology 2008;149:5922–5931.
5. de Grey A, Rae M. Ending Aging: The Rejuvenation Breakthroughs That Could Reverse Human Aging in Our Lifetime. St. Martin's Press, New York, 2007.
6. Carvey JR, Judge DS. 2000. Longevity Records: Lifespans of Mammals, Birds, Amphibians, Reptiles, and Fish. Odense University Press, Odense, Denmark, 241pp.
7. George JC, Bada J, Zeh J, Scott L, Brown SE, O'Hara T, and Suydam R. Age and growth estimates of bowhead whales (Balaena mysticetus) via aspartic acid racemization. Can J Zool 1999;77:571–580.
8. Levine, HJ. Resting heart rate and life expectancy. J Am Coll Cardiol 1997;30:1104–1106.
9. Zhang GZ, Zhang W. Heart rate, lifespan, and mortality risk. Ageing Research Reviews 2009;8:52–60.
10. Palatini P. Heart rate as a risk factor for atherosclerosis and cardiovascular mortality: The effect of antihypertensive drugs. Drugs 1999;57:713–724.
11. Harman D. The aging process. Proc Natl Acad Sci USA 1981;78:7124–7128.
12. Lloyd-Jones D, Adams R, Carnethon M, De Simone G, Ferguson TB, Flegal K,

Ford E, Furie K, Go A, Greenlund K, Haase N, Hailpern S, Ho M, Howard V, Kissela B, Kittner S, Lackland D, Lisabeth L, Marelli A, McDermott M, Meigs J, Mozaffarian D, Nichol G, O'Donnell C, Roger V, Rosamond W, Sacco R, Sorlie P, Stafford R, Steinberger J, Thom T, Wasserthiel-Smoller S, Wong N, Wylie-Rosett J, Hong Y, American Heart Association Statistics Committee and Stroke Statistics Subcommittee. Heart disease and stroke statistics– 2009 update: A report from the American Heart Association Statistics Committee and Stroke Statistics Subcommittee. Circulation 2009;119:e21–e181.

13. Berzlanovich AM, Keil W, Waldhoer T, Sim E, Fasching P, Fazeny-Dorner B. Do centenarians die healthy? An autopsy study. J Gerontol A Biol Sci Med Sci 2005;60:862–865.

14. New York Times, 1999. Accessed at www.nytimes.com/ 1999/06/09/us/anne-miller-90-first-patient-who-was-savedby-penicillin.html/.

15. Hersch EC, Merriam GR. Growth hormone (GH)-releasing hormone and GH secretagogues in normal aging: Fountain of Youth or Pool of Tantalus? Clin Interv Aging 2008;3:121–129.

16. Bluming AZ, Tavris C. Hormone replacement therapy: Real concerns and false alarms. Cancer J 2009;15:93–104.

17. Allolio B, Arlt W, Hahner S. DHEA: Why, when, and how much–DHEA replacement in adrenal insufficiency. Ann Endocrinol (Paris) 2007;68:268–273.

18. Kritz-Silverstein D, von Muhlen D, Laughlin GA, Bettencourt R. Effects of dehydroepiandrosterone supplementation on cognitive function and quality of life: The DHEA and Well-Ness (DAWN) Trial. J Am Geriatr Soc 2008;56: 1292–1298.

19. Beg S, Al-Khoury L, Cunningham GR. Testosterone replacement in men. Curr Opin Endocrinol Diabetes Obes 2008;15:364–370.

20. Borst SE, Mulligan T. Testosterone replacement therapy for older men. Clin Interv Aging 2007;2:561–566.

21. Mannino DM, Davis KJ. Lung function decline and outcomes in an elderly population. Thorax 2006;61:472–477.

22. Schunemann HJ, Dorn J, Grant BJ, Winkelstein W Jr, Trevisan M. Pulmonary function is a long-term predictor of mortality in the general population: 29-year follow-up of the Buffalo Health Study. Chest 2000;118:656–664.

23. Sin DD, Wu L, Man SF. The relationship between reduced lung function and cardiovascular mortality: A population- based study and a systematic review of the literature. Chest 2005;127:1952–1959.

24. Ryan G, Knuiman MW, Divitini ML, James A, Musk AW, Bartholomew HC. Decline in lung function and mortality: The Busselton Health Study. J Epidemiol Community Health 1999;53:230–234.

25. Alexeff SE, Litonjua AA, Sparrow D, Vokonas PS, Schwartz J. Statin use reduces decline in lung function, VA Normative Aging Study. Am J Respir Crit Care Med 2007;176:742–747.

26. Bell B, Rose C, Damon A. The Normative Aging Study: An interdisciplinary and longitudinal study of health and aging. Aging Hum Dev 1972;3:4–17.
27. Montecucco F, Mach F. Update on statin-mediated antiinflammatory activities in atherosclerosis. Semin Immunopathol 2009;31:127–142.
28. Narkar VA, Downes M, Yu RT, Embler E, Wang YX, Banayo E, Mihaylova MM, Nelson MC, Zou Y, Juguilon H, Kang H, Shaw RJ, Evans RM. AMPK and PPARdelta agonists are exercise mimetics. Cell 2008;134:405–415.
29. Agrawal S, Khan F. Human genetic variation and personalized medicine. Indian J Physiol Pharmacol 2007;51: 7–28.
30. Bottinger EP. Foundations, promises and uncertainties of personalized medicine. Mt Sinai J Med 2007;74:15–21.
31. Jain KK. Challenges of drug discovery for personalized medicine. Curr Opin Mol Ther 2006;8:487–492.
32. Subbiah MT. Nutrigenetics and nutraceuticals: the next wave riding on personalized medicine. Transl Res 2007;149: 55–61.
33. Schurgers LJ, Spronk HM, Soute BA, Schiffers PM, DeMey JG, Vermeer C. Regression of warfarin-induced medial elastocalcinosis by high intake of vitamin K in rats. Blood 2007;109:2823–2831.
34. Jain KS, Kathiravan MK, Somani RS, Shishoo CJ. The biology and chemistry of hyperlipidemia. Bioorg Med Chem 2007;15:4674–4699.
35. Prinz V, Endres M. The acute (cerebro)vascular effects of statins. Anesth Analg 2009;109:572–584.
36. Athyros VG, Kakafika AI, Tziomalos K, Karagiannis A, Mikhailidis DP. Pleiotropic effects of statins—clinical evidence. Curr Pharm Des 2009;15:479–489.
37. Mori TA, Bao DQ, Burke V, Puddey IB, Beilin LJ. Docosahexaenoic acid but not eicosapentaenoic acid lowers ambulatory blood pressure and heart rate in humans. Hypertension 1999; 34:253–260.
38. Flossmann E, Rothwell PM, British Doctors Aspirin Trial and the UK-TIA Aspirin Trial. Effect of aspirin on longterm risk of colorectal cancer: consistentevidence from randomised and observational studies. Lancet 2007;369: 1603–1613.
39. Grau MV, Sandler RS, McKeown-Eyssen G, Bresalier RS, Haile RW, Barry EL, Ahnen DJ, Gui J, Summers RW, Baron JA. Nonsteroidal anti-inflammatory drug use after 3 years of aspirin use and colorectal adenoma risk: Observational follow-up of a randomized study. J Natl Cancer Inst 2009; 101:267–276.
40. Cole BF, Logan RF, Halabi S, Benamouzig R, Sandler RS, Grainge MJ, Chaussade S, Baron JA. Aspirin for the chemoprevention of colorectal adenomas: Meta-analysis of the randomized trials. J Natl Cancer Inst 2009;101:256–266.
41. Grewal SS. Insulin/TOR signaling in growth and homeostasis: a view from the fly world. Int J Biochem Cell Biol 2009; 41:1006–1010.
42. Holloszy JO, Fontana L. Caloric restriction in humans. Exp Gerontol 2007;42:709–712.
43. Harrison DE, Strong R, Sharp ZD, Nelson JF, Astle CM, Flurkey K, Nadon NL,

Wilkinson JE, Frenkel K, Carter CS, Pahor M, Javors MA, Fernandez E, Miller RA. Rapamycin fed late in life extends lifespan in genetically heterogeneous mice. Nature 2009;460:392–395.

44. Baur JA, Pearson KJ, Price NL, Jamieson HA, Lerin C, Kalra A, Prabhu VV, Allard JS, Lopez-Lluch G, Lewis K, Pistell PJ, Poosala S, Becker KG, Boss O, Gwinn D, Wang M, Ramaswamy S, Fishbein KW, Spencer RG, Lakatta EG, Le Couteur D, Shaw RJ, Navas P, Puigserver P, Ingram DK, de Cabo R, Sinclair DA. Resveratrol improves health and survival of mice on a high-calorie diet. Nature 2006;444:337–342.

45. Lagouge M, Argmann C, Gerhart-Hines Z, Meziane H, Lerin C, Daussin F, Messadeq N, Milne J, Lambert P, Elliott P, Geny B, Laakso M, Puigserver P, Auwerx J. Resveratrol improves mitochondrial function and protects against metabolic disease by activating SIRT1 and PGC-1alpha. Cell 2006;127:1109–1122.

47. Hawley JA, Holloszy JO. Exercise: It's the real thing! Nutr Rev 2009;67:172–178.

48. Chalmers JP, Arnolda LF. Lowering blood pressure in 2003. Med J Aust 2003;179:306312.

49. Pan A, Yu D, Demark-Wahnefried W, Franco OH, Lin X. Meta-analysis of the effects of flaxseed interventions on blood lipids. Am J Clin Nutr 2009;90:288–297.

50. Kontogianni MD, Panagiotakos DB, Chrysohoou C, Pitsavos C, Zampelas A, Stefanadis C. The impact of olive oil consumption pattern on the risk of acute coronary syndromes: The CARDIO2000 case-control study. Clin Cardiol 2007;30:125–129.

51. Carlson LA. Nicotinic acid and other therapies for raising high-density lipoprotein. Curr Opin Cardiol 2006;21: 336–344.

52. Bosi E. Metformin—the gold standard in type 2 diabetes: what does the evidence tell us? Diabetes Obes Metab 2009;11(Suppl 2):3–8.

53. Lilly M, Godwin M. Treating prediabetes with metformin: Systematic review and meta-analysis. Can Fam Physician 2009;55:363–369.

54. Sadruddin S, Arora R. Resveratrol: Biologic and therapeutic implications. J Cardiometab Syndr 2009;4:102–106.

55. Nam NH. Naturally occurring NF-kappaB inhibitors. Mini Rev Med Chem 2006;6:945–951.

56. He C, Klionsky DJ. Regulation mechanisms and signaling pathways of autophagy. Annu Rev Genet 2009;43:67–93.

57. Salminen A, Kaarniranta K. Regulation of the aging process by autophagy. Trends Mol Med 2009;15:217–224.

58. Salminen A, Kaarniranta K. NF-kappaB signaling in the aging process. J Clin Immunol 2009;29:397–405.

59. Susic D, Varagic J, Ahn J, Frohlich ED. Collagen cross-link breakers: A beginning of a new era in the treatment of cardiovascular changes associated with aging, diabetes, and hypertension. Curr Drug Targets Cardiovasc Haematol

Disord. 2004;4:97–101.

60. Russo GL. Ins and outs of dietary phytochemicals in cancer chemoprevention. Biochem Pharmacol 2007;74: 533–544.

61. Perret-Guillaume C, Joly L, Jankowski P, Benetos A. Benefits of the RAS blockade: Clinical evidence before the ONTAR- GET study. J Hypertens 2009;27(Suppl 2):S3–S7.

62. Wang M, Takagi G, Asai K, Resuello RG, Natividad FF, Vatner DE, *et al*. Aging increases aortic MMP-2 activity and angiotensin II in non-human primates. Hypertension 2003; 41:1308–1316.

63. Williams B, Lacy PS, Thom SM, Cruickshank K, Stanton A, Collier D, *et al.*, CAFE Investigators. Anglo-Scandinavian Cardiac Outcomes Trial Investigators; CAFE Steering Committee and Writing Committee. Differential impact of blood pressure-lowering drugs on central aortic pressure and clinical outcomes: Principal results of the Conduit Artery Function Evaluation (CAFE) study. Circulation 2006; 113:1213–1225.

64. Cockcroft JR. ACE inhibition in hypertension: focus on perindopril. Am J Cardiovasc Drugs 2007;7:303–317.

65. Mallareddy M, Parikh CR, Peixoto AJ. Effect of angiotensin-converting enzyme inhibitors on arterial stiffness in hypertension: Systematic review and meta-analysis. J Clin Hypertens 2006;8:398–403.

66. Shargorodsky M, Leibovitz E, Lubimov L, Gavish D, Zimlichman R. Prolonged treatment with the AT1 receptor blocker, valsartan, increases small and large artery compliance in uncomplicated essential hypertension. Am J Hypertens 2002;15:1087–1091.

67. Benetos A, Levy B, Lacolley P, Taillard F, Duriez M, Safar ME. Role of angiotensin II and bradykinin on aortic collagen following converting enzyme inhibition in spontaneously hypertensive rats. Arterioscler Thromb Vasc Biol 1997;17:3196–3201.

68. Labat C, Lacolley P, Lajemi M, De Gasparo M, Safar ME, Benetos A. Effects of valsartan on mechanical properties of the carotid artery in SHR under high salt diet. Hypertension 2001;38:439–443.

69. Ceconi C, Fox KM, Remme WJ, Simoons ML, Bertrand M, Parrinello G, *et al*. ACE inhibition with perindopril and endothelial function. Results of a substudy of the EUROPA study: PERTINENT. Cardiovasc Res 2007;73:237–246.

70. Galderisi M, de Divitiis O. Risk factor-induced cardiovascular remodeling and the effects of angiotensin-converting enzyme inhibitors. J Cardiovasc Pharmacol 2008;51:523–531 71.Hayek T, Attias J, Coleman R, Brodsky S, Smith J, Breslow JL *et al*. The angiotensin-converting enzyme inhibitor, fosinopril, and the angiotensin II receptor antagonist, losartan, inhibit LDL oxidation and attenuate atherosclerosis independent of lowering blood pressure in apo- lipoprotein E deficient mice. Cardiovasc Res 1999;44:579–587.

72. Hayek T, Attias J, Smith J, Breslow JL, Keidar S. Anti-atherosclerotic and

antioxidative effects of captopril in apolipoproteinE-deficient mice. J Cardiovasc Pharmacol 1998;31:540–544.

73 Brilla CG, Rupp H, Maisch B. Effects of ACE inhibition versus non-ACE inhibitor antihypertensive treatment on myocardial fibrosis in patients with arterial hypertension. Retrospective analysis of 120 patients with left ventricular endomyocardial biopsies. Herz 2003;28:744–753.

74. Klingbeil AU, Schneider M, Martus P, Messerli FH, Schmieder RE. A meta-analysisof the effects of treatment on left ventricular mass in essential hypertension. Am J Med 2003;115:41–46.

75. Lonn E, Shaikholeslami R, Yi Q, Bosch J, Sullivan B, Tanser P, et al. Effects of ramipril on left ventricular mass and function in cardiovascular patients with controlled blood pressure and with preserved left ventricular ejection fraction: A substudy of the Heart Outcomes Prevention Evaluation (HOPE) Trial. J Am Coll Cardiol 2004;43:2200–2206.

76. Healey JS, Baranchuk A, Crystal E, Morillo CA, Garfinkle M, Yusuf S. Prevention of atrial fibrillation with angiotensin- converting enzyme inhibitors and angiotensin receptor blockers: A meta-analysis. J Am Coll Cardiol 2005;45: 1832–1839.

76. Smith JJ, Kenney RD, Gagne DJ, Frushour BP, Ladd W, Galonek HL, Israelian K, Song J, Razvadauskaite G, Lynch AV, Carney DP, Johnson RJ, Lavu S, Iffland A, Elliott PJ, Lambert PD, Elliston KO, Jirousek MR, Milne JC, Boss O. Small molecule activators of SIRT1 replicate signaling pathways triggered by calorie restriction in vivo. BMC Syst Biol 2009;3:31.

77. Ehrlich JR, Hohnloser SH, Nattel S. Role of angiotensin system and effects of its inhibition in atrial fibrillation: clinical and experimental evidence. Eur Heart J 2006;27: 512–518.

78. Wachtell K, Lehto M, Gerdts E, Olsen MH, Hornestam B, Dahlof B, et al. Angiotensin II receptor blockade reduces new-onset atrial fibrillation and subsequent stroke compared to atenolol: the Losartan Intervention For End Point Reduction in Hypertension (LIFE) study. J Am Coll Cardiol 2005;45:712–719.

79. Andrews R, Brown DL. Effect of inhibition of the renin–angiotensin system on development of type 2 diabetes mellitus (meta-analysis of randomized trials). Am J Cardiol 2007;99:1006–1012.

80. Tikellis C, Cooper ME, Thomas MC. Role of the renin– angiotensin system in the endocrine pancreas: implications for the development of diabetes. Int J Biochem Cell Biol 2006;38:737–751.

81. Benson P, Pravenec M, et al. Identification of telmisartan as a unique angiotensin II receptor antagonist with selective PPAR- gamma-modulating activity. Hypertension 2004;43:993–1002.

82. Strippoli GF, Craig M, Schena FP, Craig JC. Antihypertensive agents for primary prevention of diabetic nephropathy. J Am Soc Nephrol 2005;16:3081–3091.

83. Dorffel Y, Latsch C, Stuhlmuller B, Schreiber S, Scholze S, Burmester GR, Scholze J. Preactivated peripheral blood monocytes in patients with essential hypertension. Hypertension 1999;34:113–117.

84. Platten M, Youssef S, Hur EM, Ho PP, Han MH, Lanz TV, Phillips LK, Goldstein MJ, Bhat R, Raine CS, Sobel RA, Steinman L. Blocking angiotensin-converting enzyme induces potent regulatory T cells and modulates TH1- and TH17-mediated autoimmunity. Proc Natl Acad Sci USA 2009;106:14948–14953.

85. Geara AS, Azzi J, Jurewicz M, Abdi R. The renin-angiotensin system: An old, newly discovered player in immunoregulation. Transplant Rev (Orlando) 2009;23: 151–158.

86. Basso N, Cini R, Pietrelli A, Ferder L, Terragno NA, Inserra F. Protective effect of long-term angiotensin II inhibition. Am J Physiol Heart Circ Physiol 2007;293: H1351–H1358.

87. Benigni A, Corna D, Zoja C, Sonzogni A, Latini R, Salio M, Conti S, Rottoli D, Longaretti L, Cassis P, Morigi M, Coffman TM, Remuzzi G. Disruption of the Ang II type 1 receptor promotes longevity in mice. J Clin Invest 2009;119: 524–530.

88. Egami K, Murohara T, Shimada T, Sasaki K, Shintani S, Sugaya T, Ishii M, Akagi T, Ikeda H, Matsuishi T, Imaizumi T. Role of host angiotensin II type 1 receptor in tumor angiogenesis and growth. J Clin Invest 2003;112: 67–75.

89. Treuting PM, Linford NJ, Knoblaugh SE, Emond MJ, Morton JF, Martin GM, Rabinovitch PS, Ladiges WC. Reduction of age-associated pathology in old mice by over-expression of catalase in mitochondria. J Gerontol A Biol Sci Med Sci 2008;63:813–822.

90. Lipman RD, Bronson RT, Wu D, Smith DE, Prior R, Cao G, Han SN, Martin KR, Meydani SN, Meydani M. Disease incidence and longevity are unaltered by dietary antioxidant supplementation initiated during middle age in C57BL/6 mice. Mech Ageing Dev 1998;103:269–284.

91. Weindruch R, Masoro EJ. Concerns about rodent models for aging research. J Gerontol 199;46:B87–B88.

92. Dai DF, Santana LF, Vermulst M, Tomazela DM, Emond MJ, MacCoss MJ, Gollahon K, Martin GM, Loeb LA, Ladiges WC, Rabinovitch PS. Overexpression of catalase targeted to mitochondria attenuates murine cardiac aging. Circulation 2009;119:2789–27897.

93. Yao HW, Zhu JP, Zhao MH, Lu Y. Losartan attenuates bleomycin-induced pulmonary fibrosis in rats. Respiration 2006;73:236–242.

94. Krueger T, Westenfeld R, Schurgers L, Brandenburg V. Coagulation meets calcification: The vitamin K system. Int J Artif Organs 2009;32:67–74.

95. Cranenburg EC, Brandenburg VM, Vermeer C, Stenger M, Muhlenbruch G, Mahnken AH, Gladziwa U, Ketteler M, Schurgers LJ. Uncarboxylated matrix Gla protein (ucMGP) is associated with coronary artery calcification in haemodialysis patients. Thromb Haemost 2009;101:359– 81.

96. Cranenburg EC, Vermeer C, Koos R, Boumans ML, Hackeng TM, Bouwman

FG, Kwaijtaal M, Brandenburg VM, Ketteler M, Schurgers LJ. The circulating inactive form of matrix Gla Protein (ucMGP) as a biomarker for cardiovascular calcification. J Vasc Res 2008;45:427–436.

97. Hadi HA, Mahmeed WA, Suwaidi JA, Ellahham S. Pleiotropic effects of statins in atrial fibrillation patients: the evidence. Vasc Health Risk Manag.2009;5:533–551.

98. Kurian KC, Rai P, Sankaran S, Jacob B, Chiong J, Miller AB. The effect of statins in heart failure: beyond its cholesterol- lowering effect. J Card Fail 2006;12:473–478.

99. Lavie CJ, Mehra MR. Statins and advanced heart failure— alive but barely breathing after CORONA and GISSI-HF. Congest Heart Fail 2009;15:157–158.

100. Martin JH, Mangiafico S, Kelly DJ. Role of statins in diabetes complications. Curr Diabetes Rev 2009;5:165–170.

101. Brugts JJ, Yetgin T, Hoeks SE, Gotto AM, Shepherd J, Westendorp RG, de Craen AJ, Knopp RH, Nakamura H, Ridker P, van Domburg R, Deckers JW. The benefits of statins in people without established cardiovascular disease but with cardiovascular risk factors: Meta-analysis of randomised controlled trials. BMJ 2009;338:b2376.

102. Vanschoonbeek K, Wouters K, van der Meijden PE, van Gorp PJ, Feijge MA, Herfs M, Schurgers LJ, Hofker MH, de Maat MP, Heemskerk JW. Anticoagulant effect of dietary fish oil in hyperlipidemia: a study of hepatic gene expression in APOE2 knock-in mice. Arterioscler Thromb Vasc Biol 2008; 28:2023–2029.

103. Bruunsgaard H, Pedersen BK. Age-related inflammatory cytokines and disease. Immunol Allergy Clin North Am 2003;23:15–39.

104. Khatami M. Inflammation, aging, and cancer: Tumoricidal versus tumorigenesis of immunity: A common denominator mapping chronic diseases. Cell Biochem Biophys 2009; 55:55–79.

105. Weisman D, Hakimian E, Ho GJ. Interleukins, inflammation, and mechanisms of Alzheimer's disease. Vitam Horm 2006;74:505–530.

106. Banning M. The principles of inflammation in the development of rheumatoid arthritis. Br J Nurs 2005;14: 277–283.

107. Barnes PJ. Future treatments for chronic obstructive pulmonary disease and its comorbidities. Proc Am Thorac Soc 2008;5:857–864.

108. Packard RR, Libby P. Inflammation in atherosclerosis: from vascular biology to biomarker discovery and risk prediction. Clin Chem 2008;54:24–38.

109. Bray GA, Clearfield MB, Fintel DJ, Nelinson DS. Overweight and obesity: the pathogenesis of cardiometabolic risk. Clin Cornerstone 2009;9:30–40; discussion 41–42.

110. Inagi R. Oxidative stress in cardiovascular disease: A new avenue toward future therapeutic approaches. Recent Pat Cardiovasc Drug Discov 2006;1:151–159.

111. Kang JX, Weylandt KH. Modulation of inflammatory cytokines by omega-3

fatty acids. Subcell Biochem 2008;49: 133–143.

112. Wolff T, Miller T, Ko S. Aspirin for the primary prevention of cardiovascularevents: an update of the evidence for the U.S. Preventive Services Task Force. Ann Intern Med 2009; 150:405–410.

113. Vlad SC, Miller DR, Kowall NW, Felson DT. Protective effects of NSAIDs on the development of Alzheimer's disease. Neurology 2008;70:1672–1677.

114. Szekely CA, Breitner JC, Fitzpatrick AL, Rea TD, Psaty BM, Kuller LH, Zandi PP. NSAID use and dementia risk in the Cardiovascular Health Study: Role of APOE and NSAID type. Neurology 2008;70:17–24.

115. Barzilai N, Bartke A. Biological approaches to mechanistically understand the healthy lifespan extension achieved by calorie restriction and modulation of hormones. J Gerontol A Biol Sci Med Sci 2009;64:187–191.

116. Harman D. Aging: A theory based on free radical and radiation chemistry. J Gerontol 1956;11:298–300.

117. Kamel NS, Gammack J, Cepeda O, Flaherty JH. Antioxidants and hormones as antiaging therapies: high hopes, disappointing results. Cleveland Clin J Med 2006;73:1049– 1050.

118. Fusco D, Colloca G, Lo Monaco MR, Cesari M. Effects of antioxidant supplementation on the aging process. Clin Interv Aging 2007;2:377–387.

119. Green GA. Review: Antioxidant supplements do not reduce all-cause mortality in primary or secondary prevention. Evid Based Med 2008;13:177.

120. Ristow M, Zarse K, Oberbach A, Kloting N, Birringer M, Kiehntopf M, Stumvoll M, Kahn CR, Bluher M. Antioxidants prevent health-promoting effects of physical exercise in humans. Proc Natl Acad Sci USA 2009;106:8665–8670.

121. Aigaki T, Seong KH, Matsuo T. Longevity determination genes in *Drosophila melanogaster*. Mech Ageing Dev 2002; 123:1531–1541.

122. Perez VI, Bokov A, Remmen HV, Mele J, Ran Q, Ikeno Y, Richardson A. Is the oxidative stress theory of aging dead? Biochim Biophys Acta 2009;1790:1005–1014.

123. Lapointe J, Stepanyan Z, Bigras E, Hekimi S. Reversal of the mitochondrial phenotype and slow development of oxidative biomarkers of aging in long-lived Mclk1+/- mice. J Biol Chem 2009;284:20364–20374.

123. Schriner SE, Linford NJ, Martin GM, Treuting P, Ogburn CE, Emond M, Coskun PE, Ladiges W, Wolf N, Van Remmen H, Wallace DC, Rabinovitch PS. Extension of murine lifespan by overexpression of catalase targeted to mitochondria. Science 2005;308:1909–1911.

124. Andziak B, O'Connor TP, Qi W, DeWaal EM, Pierce A, Chaudhuri AR, Van Remmen H, Buffenstein R. High oxidative damage levels in the longest-living rodent, thenaked mole-rat. Aging Cell 2006;5:463–471.

124. Perez VI, Van Remmen H, Bokov A, Epstein CJ, VijgJ, Richardson A. The overexpression of major antioxidant enzymes does not extend the lifespan of mice. Aging Cell 2009;8:73–75.

125. Van Raamsdonk JM, Hekimi S. Deletion of the mitochondrial superoxide dismutase sod-2 extends lifespan in Caenorhabditis elegans. PLoS Genet 2009;5:e1000361.

126. Fan X, Reneker LW, Obrenovich ME, Strauch C, Cheng R, Jarvis SM, Ortwerth BJ, Monnier VM. Vitamin C mediates chemical aging of lens crystallins by the Maillard reaction in a humanized mouse model. Proc Natl Acad Sci USA 2006;103:16912–16917.

127. Chan EC, Jiang F, Peshavariya HM, Dusting GJ. Regulation of cell proliferation by NADPH oxidase-mediated signaling: Potential roles in tissue repair, regenerative medicine and tissue engineering. Pharmacol Ther 2009;122: 97–108.

130. Suzuki K. Anti-oxidants for therapeutic use: Why are only a few drugs in clinical use? Adv Drug Deliv Rev 2009; 61: 287–289.

131. Groban L, Pailes NA, Bennett CD, Carter CS, Chappell MC, Kitzman DW, Sonntag WE. Growth hormone replacement attenuates diastolic dysfunction and cardiac angiotensin II expression in senescent rats. J Gerontol A Biol Sci Med Sci 2006;61:28–35.

132. Zou XJ, Yang L, Yao SL. Propofol depresses angiotensin II-induced cardiomyocyte hypertrophy in vitro. Exp Biol Med (Maywood) 2008;233:200–208.

133. Bao W, Behm DJ, Nerurkar SS, Ao Z, Bentley R, Mirabile RC, Johns DG, Woods TN, Doe CP, Coatney RW, Ohlstein JF, Douglas SA, Willette RN, Yue TL. Effects of p38 MAPK Inhibitor on angiotensin II-dependent hypertension, organ damage, and superoxide anion production. J Cardiovasc Pharmacol 2007;49:362–368.

134. Alcendor RR, Gao S, Zhai P, Zablocki D, Holle E, Yu X, Tian B, Wagner T, Vatner SF, Sadoshima J. Sirt1 regulates aging and resistance to oxidative stress in the heart. Circ Res 2007;100:1512–1521.

135. Herbert KE, Mistry Y, Hastings R, Poolman T, Niklason L, Williams B. Angiotensin II-mediated oxidative DNA damage accelerates cellular senescence in cultured human vascular smooth muscle cells via telomere-dependent and independent pathways. Circ Res 2008;102:201–208.

136. Terman A. Garbage catastrophe theory of aging: Imperfect removal of oxidative damage? Redox Rep 2001;6:15–26.

137. Budovskaya YV, Wu K, Southworth LK, Jiang M, Tedesco P, Johnson TE, Kim SK. An elt-3/elt-5/elt-6 GATA transcription circuit guides aging in *C. elegans*. Cell 2008;134: 291–303.

138. Raices M, Maruyama H, Dillin A, Karlseder J. Uncoupling of longevity and telomere length in *C. elegans*. PLoS Genet 2005;1:e30. Erratum in: PloS Genet 2005;1:e81.

139. Stripp BR, Reynolds SD. Maintenance and repair of the bronchiolar epithelium. Proc Am Thorac Soc 2008;5:328– 333.

140. Rawlins EL, Hogan BL. Ciliated epithelial cell lifespan in the mouse trachea

and lung. Am J Physiol 2008;295:L231– L234.

141. Giangreco A, Arwert EN, Rosewell IR, Snyder J, Watt FM, Stripp BR. Stem cellsare dispensable for lung homeostasis but restore airways after injury. Proc Natl Acad Sci USA 2009;106:9286–9291.

142. Dor Y, Brown J, Martinez OI, Melton DA. Adult pancreatic beta-cells are formed by self-duplication rather than stemcell differentiation. Nature 2004;429:41–46.

143. Bergmann O, Bhardwaj RD, Bernard S, Zdunek S, Barnabe Heider F, Walsh S, Zupicich J, Alkass K, Buchholz BA, Druid H, Jovinge S, Frise n J. Evidence for cardiomyocyte renewal in humans. Science 2009;324:98–102.

144. Martinez DE. Mortality patterns suggest lack of senescence in hydra. Exp Gerontol 1998;33:217–225.

145. Newmark PA, Sanchez Alvarado A. Not your father's planarian: A classic model enters the era of functional genomics. Nat Rev Genet 2002;3:210–219.

146. Rajnoch C, Ferguson S, Metcalfe AD, Herrick SE, Willis HS, Ferguson MW. Regeneration of the ear after wounding in different mouse strains is dependent on the severity of wound trauma. Dev Dyn 2003;226:388–397.

147. Reines B, Cheng LI, Matzinger P. Unexpected regeneration in middle-aged mice. Rejuvenation Res 2009;12:45–52.

148. Hsieh PC, Segers VF, Davis ME, Macgillivray C, Gannon J, Molkentin JD, Robbins J, Lee RT. Evidence from a genetic fate-mapping study that stem cells refresh adult mammalian cardiomyocytes after injury. Nat Med 2007;13: 970–974.

149. Bersell K, Arab S, Haring B, Ku hn B. Neuregulin1/ErbB4 signaling induces cardiomyocyte proliferation and repair of heart injury. Cell 2009;138:257–270.

150. Chen S, Zhang Q, Wu X, Schultz PG, Ding S. Dedifferentiation of lineage-committed cells by a small molecule. J Am Chem Soc 2004;126:410–411.

151. Lee RJ, Fang Q, Davol PA, Gu Y, Sievers RE, Grabert RC, Gall JM, Tsang E, Yee MS, Fok H, Huang NF, Padbury JF, Larrick JW, Lum LG. Antibody targeting of stem cells to infarcted myocardium. Stem Cells 2007;25:712–717.

152. Takahashi K, Yamanaka S. Induction of pluripotent stem cells from mouse embryonic and adult fibroblast cultures by defined factors. Cell 2006;126:663–676.

153. Takahashi K, Tanabe K, Ohnuki M, Narita M, Ichisaka T, Tomoda K, Yamanaka S. Induction of pluripotent stem cells from adult human fibroblasts by defined factors. Cell 2007;131:861–872.

154. Zhou H, Wu S, Joo JY, Zhu S, Han DW, Lin T, Trauger S, Bien G, Yao S, Zhu Y, Siuzdak G, Scholer HR, Duan L, Ding S. Generation of induced pluripotent stem cells using recombinant proteins. Cell Stem Cell 2009;4:381–384. Erratum in: Cell Stem Cell 2009;4:581.

155. Shi Y, Desponts C, Do JT, Hahm HS, Scholer HR, Ding S. Induction of pluripotent stem cells from mouse embryonic fibroblasts by Oct4 and Klf4 with small-molecule compounds. Cell Stem Cell 2008;3:568–574.

156. Lyssiotis CA, Foreman RK, Staerk J, Garcia M, Mathur D, Markoulaki S, Hanna J, Lairson LL, Charette BD, Bouchez LC, Bollong M, Kunick C, Brinker A, Cho CY, Schultz PG, Jaenisch R. Reprogramming of murine fibroblasts to induced pluripotent stem cells with chemical complementation of Klf4. Proc Natl Acad Sci USA 2009;106:8912–8917.

157. Zhang J, Wilson GF, Soerens AG, Koonce CH, Yu J, Palecek SP, Thomson JA, Kamp TJ. Functional cardiomyocytes derived from human induced pluripotent stem cells. Circ Res. 2009;104:e30–e41.

158. Menasche P. Skeletal myoblasts as a therapeutic agent. Prog Cardiovasc Dis 2007;50:7–17.

159. Laflamme MA, Chen KY, Naumova AV, Muskheli V, Fugate JA, Dupras SK, Reinecke H, Xu C, Hassanipour M, Police S, O'Sullivan C, Collins L, Chen Y, Minami E, Gill EA, Ueno S, Yuan C, Gold J, Murry CE. Cardiomyocytes derived from human embryonic stem cells in pro-survival factors enhance function of infarcted rat hearts. Nat Biotechnol 2007;25:1015–1024.

160. Kobayashi N. Life support of artificial liver: Development of a bioartificial liver to treat liver failure. J Hepatobiliary Pancreat Surg 2009;16:113–117.

161. Hayashida K, Kanda K, Oie T, Okamoto Y, Sakai O, Watanabe T, Ishibashi-Ueda H, Onoyama M, Tajikawa T, Ohba K, Yaku H, Nakayama Y. "*In vivo* tissue-engineered" valved conduit with designed molds and laser processed scaffold. J Cardiovasc Nurs 2008;23:61–64.

162. Morris RJ. Total artificial heart—concepts and clinical use. Semin Thorac Cardiovasc Surg 2008;20:247–254.

163. Mountis MM, Starling RC. Management of left ventricular assist devices after surgery: Bridge, destination, and recovery. Curr Opin Cardiol 2009;24:252–256.

164. Quiroga QR, Kraskov A, Koch C, Fried I. Explicit encoding of multimodal percepts by single neurons in the human brain. Curr Biol 2009;19:1308–1313.

165. Lee E, Son H. Adult hippocampal neurogenesis and related neurotrophic factors. BMB Rep 2009;42:239–244.

166. Kim SJ, Son TG, Park HR, Park M, Kim MS, Kim HS, Chung HY, Mattson MP, Lee J. Curcumin stimulates proliferation of embryonic neural progenitor cells and neurogenesis in the adult hippocampus. J Biol Chem 2008;283: 14497–14505.

167. van Praag H. Exercise and the brain: something to chew on. Trends Neurosci 2009;32:283–290.

INDEX

ACE 4, 9pp., 13, 15, 18, 20, 27, 29, 31pp., 35p., 38p., 42, 44, 54pp., 60, 66, 68, 74, 76p., 80p., 91, 114, 122, 131p., 134pp., 140, 142p., 145pp., 152, 155, 157p., 162pp., 166p., 172, 177pp., 183, 188p., 193, 201pp., 207pp., 218, 221, 223p., 226pp., 231p., 234pp., 238, 241, 243p., 248p., 251p., 255p., 258, 262p., 266, 268pp., 281p., 285p., 310pp.
acetaminophen..........157, 162, 266
acetyltransferases......................201
actigraphy................................186
actin......42, 53, 87, 107, 109, 203, 214, 223, 253
actinomycin..............................203
acyl-coa....................................69
adenomas.................................279
adenosine 23p., 27, 62, 66, 93, 97, 99, 106, 128, 134, 138, 158, 184, 189, 223
adhesion....................164, 172, 261
adipocytes.........................224, 271
adiponectin.......................138, 161
adipose.....75, 108, 114, 144, 262, 272
adrenal.................................32, 278
adrenergic. 23, 32, 70, 97pp., 175, 181p., 187p., 205
aerobic.....29p., 61, 132, 155, 169, 172, 275
aging..
    Ageing....56p., 68, 94p., 113p., 130p., 172, 198pp., 218p., 232p., 235, 242, 277, 283, 285
agonists24, 27, 101, 106, 134, 140, 143p., 168, 181, 187, 207, 231p., 279
AICAR..
    aicar....27, 67, 70, 93, 112, 134, 141
airways..............................165, 287
aldosterone..............................239
alopecia....................................88
alpha-synuclein...................31, 176
Alzheimer's disease.......................
    Alzheimer's.........17p., 40, 58p., 189p., 209, 284
ambulatory..............................279
aminoimidazole............67, 93, 112
aminopeptidases.......................176
amphibians..............................277
AMPK.....20, 24, 66p., 69, 96, 106, 112, 115, 128, 144, 158, 164, 173, 235, 279
amyloid.......17p., 31, 49p., 54pp., 58p., 90p., 175p., 182, 185, 189p., 208
angiotensin.....152, 188, 231, 239, 257p., 266p., 281pp., 286
angiotensin-converting...152, 188, 281, 283
anti-aging 5, 18, 22p., 25, 28p., 35, 37, 39, 84p., 93, 111, 155, 157, 210, 218, 238, 241
anti-atherosclerotic...........258, 281
anti-autoimmune.......................261

antiaging......4, 50, 92, 111, 113p., 256, 285
antibiotics....................147pp., 249
antibodies...........44, 196, 272, 310
   Antibody.......127, 196, 287, 310
anticoagulant...........................284
anticonvulsant..........................205
antigen.......................93, 127, 272
antihypertensive........190, 277, 282
antihypertrophic......................240
antioxidants.................................
   anti-oxidant......155pp., 161pp., 168pp., 224, 226pp., 286
   antioxidant13, 16, 29p., 38, 155, 161, 169, 171p., 234, 263pp., 283, 285
   antioxidative........................282
antioxidative............................282
antiproliferative.........................93
antitrypsin..................................55
aortic................................240, 281
apigenin.....................34, 197, 200
APOE 18, 33, 136, 148, 150, 175p., 186pp., 190, 284p.
apoptosis...................132, 163, 224
aptamers..................................239
Aquaporin................................178
arachidonic..............................266
ARBs........................152, 258, 262
   ARBs...7, 24, 27, 55, 57, 62, 67, 93, 106, 108, 112p., 122, 134, 141, 152, 161, 172, 176, 188, 200, 215, 219, 233, 247, 258, 261p., 267p., 271, 274, 283, 286
   Losartan...........231, 258, 281pp.

   Telmisartan............................282
   Valsartan...............................281
arginine....................................151
arrhythmia.................160, 262, 273
arrhythmias......................160, 273
artery...................147, 271, 281, 283
arthralgia...................................92
arthritis....160, 169, 174, 262, 266, 284
artifactual................................239
ascorbic...........................173p., 266
aspartate..................................210
aspartic.....................218, 248, 277
aspirin..............12p., 262, 279, 285
astragalus-derived....21, 71, 77, 79
astrocytes..........................32, 178
astroglial..........................175, 182
asymmetric..............................277
atenolol....................................282
atheromas...............................146
atherosclerosis......22, 84, 88, 131, 133, 145p., 148pp., 154, 157, 164, 172p., 261p., 276p., 279, 281, 284
ATPase.....................................238
atrial.................238, 261, 282, 284
autocrine.................................135
autoimmunity..........................283
autophagasomes........................90
autophagy.22p., 48, 52, 55, 58, 61, 84p., 87pp., 93p., 96, 106p., 113, 115, 119, 263, 268, 280
b-d-ribofuranoside................67, 93
bacteria......................42, 150, 153
Bacteroides..............................150
Bafilomycin................................90
basophils.................................200

bdnf...............35p., 200pp., 207pp.
beta-catenin........21, 71, 76pp., 82
beta-cells...................................287
beta-klotho..............................135
betaine.....................................147
binucleated..............................271
bioartificial..............................288
biochemical.....11, 13pp., 26, 100, 153, 216, 222, 247, 255
biodevice..................................274
bioenergetics...........................133
bioengineering.................252, 274
    bioengineered...................273p.
bioflavonoid............................264
biologic.........6, 8pp., 12p., 18, 30, 39pp., 170p., 210, 213p., 217p., 252, 255, 274, 280, 285, 313
biomarker 5, 26, 28, 36p., 49, 116, 125, 129pp., 145, 152p., 210pp., 218, 235, 276, 284p.
biomarkers........26, 36p., 116, 125, 129pp., 210pp., 276, 285
    biomarker....5, 26, 28, 36p., 49, 116, 125, 129pp., 145, 152p., 210pp., 218, 235, 276, 284p.
biomaterials.............................274
biomechanical..................247, 274
biomolecules 14, 16, 31p., 44, 156, 165, 175pp., 181, 184, 222, 228
biopsies..........................161p., 282
biosynthesis.......................132, 253
birds...277
bispecific..................................272
blastema...........................43, 270
bleomycin-induced...........258, 283
blocker100, 152, 187p., 258, 281p.

bone-marrow............................272
bowhead............................248, 277
bradykinin................................281
brain..5, 13p., 30pp., 58p., 70, 74, 91, 95, 100p., 104, 121, 143, 175pp., 184p., 187pp., 192, 195, 197, 200pp., 204, 209, 219p., 238, 264, 267, 270, 275, 288
BrdU...195
breakthrough 10, 16, 68, 182, 272, 277
breathing...........................169, 284
bromodeoxyuridine..................195
bronchiolar..............................286
bronchioles..............................269
bronchoconstriction.........165, 170, 173p.
bronchoconstrictors..................166
bronchospasm..........................173
butyrate...27, 54, 134, 140p., 143, 203
c-Myc..272
C-reactive..........................125, 261
c-Src...78
Caaenorhabditis........................202
calcification....................261, 283p.
calmodulin..........................67, 203
calnexin......................................52
caloric restriction.......25p., 37, 125, 130pp., 138, 144, 157, 171, 210, 223, 234, 263, 279
calorie....26, 88, 117, 124, 126pp., 131, 133, 158, 234p., 280, 282, 285
calpain......................................176
CaMKII............................203p., 208

cAMP 9, 23, 36, 57, 68, 70, 80, 94, 97, 99pp., 104, 194, 199pp., 204p., 208p., 219, 275, 288
cancer...9pp., 16, 20p., 26, 30, 42, 52, 62, 71pp., 78pp., 92pp., 106, 111pp., 115p., 120p., 124, 127p., 130p., 141, 143, 155, 157, 160, 167pp., 174, 205, 214, 216, 220, 235, 247, 255, 258, 262, 266, 270, 278p., 281, 284, 310p.
    neoplasms..............................255
    tumor...10, 21, 23, 25, 42p., 52, 62, 65, 72, 76, 78pp., 92p., 125, 132, 144, 168, 216, 255, 258, 262, 283p.
        tumoricidal..............................284
        tumorigenesis..52, 65, 72, 78p., 168, 284
captopril.......................................282
cardiomyopathy.........................62p.
carnitine........28, 40, 145p., 149p., 152pp.
carotid..281
cascade................................203, 266
caspases......................................176
catabolize......................................55
catalase119, 224, 237, 241p., 265, 267, 283, 285
catalytically...........................76, 226
cataract..............................218, 266
catechins..59
catheterization............................274
    catheter.....................................274
cats..248
CBP-dependent..........................208
celastrol..................................55, 58

cell cycle..43, 94p., 113, 132, 144, 214, 227, 232, 271
cell differentiation..........................
    differentiation....5, 14p., 37, 39, 93, 96, 141, 144, 146, 154, 191, 193, 196, 199, 202, 210pp., 217p., 223p., 226, 229, 236pp., 242p., 272, 287
centenarians...........220, 249, 277p.
cerevisiae..............................85, 106
cetaceans....................................248
chemokines........................195, 199
chemoprevention...............279, 281
chemotactic.................................163
chemotherapy.............................255
    chemoprevention..........279, 281
cholesterol..28, 92, 111, 121, 125, 129, 136, 140, 145p., 148, 151p., 163, 176, 253, 284
chondrocytes..............................271
cigarettes....................................253
ciprofloxacin...............................149
circuit..95, 99, 161, 163, 212, 216, 252, 286
circulating...34, 39, 143, 172, 174, 192, 209, 237pp., 242, 258, 284
clonidine.....................................105
Coagulation........................274, 283
coenzyme Q................................265
colorectal....................................279
coronary.........131, 146p., 280, 283
CpG....................37, 210pp., 219p.
creb.........................67, 203, 208p.
crystallins....................................286
curcumin..17p., 34, 49, 55, 58, 93, 96, 197, 200, 217, 275, 288

cycloastragenol...................77p., 81
cyclosporine................................93
cysteine....29, 38p., 155, 157, 162, 172, 227, 266
cytokines. 195, 262, 266, 277, 284, 310
cytokinesis................................277
Cytomegalovirus....................78, 93
cytosines..................................211
cytoskeleton........................87, 107
d-aspartate..............................210
damage....2, 4, 9, 14pp., 34, 37p., 42, 46, 48, 61, 63, 72, 89, 100, 119, 126, 140, 156p., 160pp., 172, 176, 192, 194, 205, 210, 213, 216p., 221pp., 228p., 231p., 234pp., 246pp., 258, 262pp., 270, 272p., 277, 285p.
deaminase......................20, 67, 69
degenerative......4p., 11, 17, 22p., 30pp., 48p., 56, 84, 90, 92, 94, 97p., 114, 175p., 184p., 187p., 205, 209, 249, 256
dehydroepiandrosterone..........278
dehydrogenase.................123, 224
dementia..........58, 179, 185p., 285
dendritic....93, 96, 99, 199, 201pp.
Deoxygedunin.............36, 207, 209
desmosterol.............................151
deuterated.......................147, 149
development.9, 14, 18, 21, 23, 28, 37, 39, 43p., 56, 68, 71, 73, 76p., 79, 84, 88, 94, 96, 113, 115, 129, 133, 139, 145p., 148, 154, 156, 160, 167, 187, 190, 198, 207, 210, 215, 218, 222, 229, 233, 236, 238, 241p., 244, 246, 255pp., 264, 266, 268, 270pp., 276, 282, 284p., 288, 310p., 313
dextran....................................179
DHEA...............................252, 278
diabetes24, 40, 54, 58, 94, 113pp., 121p., 124, 126, 128, 131, 139, 141, 143p., 157, 161, 232, 262, 276, 278, 280, 282, 284
  diabetic...93, 144, 169, 258, 282
diastolic. 126, 129, 237p., 242, 286
dibromo.....................................54
dietary restriction...4, 25p., 34, 40, 50p., 91, 108p., 116p., 119, 130p., 133p., 142p., 171, 197, 200, 217
diethylstilbestrol.......................277
diets....25, 28, 116, 125, 129, 133, 145p., 148, 231, 264
Diltiazem...................................55
dismutase..38, 119, 160, 162, 221, 224, 265, 286
DNA...5, 17, 19, 36pp., 44, 61, 63, 80, 89, 163, 210pp., 218pp., 224, 228p., 234p., 267, 271, 286
docosahexaenoic..............262, 279
dopaminergic.............................91
doublecortin.............................195
drosophila..51, 57p., 85, 106, 111, 116p., 120, 130p., 141, 144, 158, 171, 189, 211, 233, 263, 265, 285
drugs 9p., 18, 20p., 27, 37, 40, 53, 56, 66, 68, 71, 76, 78p., 92, 94, 100p., 104, 106, 112p., 119, 127, 134, 187, 190, 209p., 217, 229, 241, 253, 258, 261p., 271, 273, 276p., 281, 286

dyslipidemia..............................140
Dysregulation.....23, 84p., 89, 104, 109, 240
e-selectin...................................172
edaravone..................................266
eicosapentaenoic.....................279
elastocalcinosis.................261, 279
elegans..36, 48p., 51p., 55pp., 63, 66, 85, 95, 106p., 114, 120, 158, 171, 201p., 208, 211, 232p., 263, 265, 268p., 286
ELISAs..164
embryonic.......109, 200, 272, 277, 287p.
enapril..258
endocannabinoid......................209
endotherm...........................52, 57
enzymes......55, 62, 108, 112, 119, 157, 160, 162, 176, 226, 265, 285
eosinophils........................198, 200
eotaxin...............34, 192, 194, 200
epigallocatechin....................56, 59
epigenetics.............5, 15, 191, 211
epithelial...............43, 73, 269, 286
estrogen....................................252
euglycemia................................136
everolimus.....................92, 96, 114
evolution..8, 14, 26, 32, 41pp., 45, 72, 85, 107, 119, 154, 160, 165, 170, 193, 205, 233p., 244, 256p., 263, 266, 276
exercise...5, 10, 12p., 16, 20, 29p., 33p., 36, 38, 40, 65, 67pp., 128pp., 141, 144, 152, 155p., 160pp., 169pp., 187, 197, 200, 207, 209, 231, 235, 244, 248, 255, 262, 265, 275p., 279p., 285, 288
eyes..98
fate-mapping...........................287
fenofibrate...................27, 134, 140
fibrillation..................262, 282, 284
fibroblasts..........78, 88, 218, 287p.
fibroconnective........................249
fibrosis.....53p., 58, 199, 252, 258, 266p., 282p.
  fibrotic...................................267
FITC-dextran.............................179
flavanol..59
flavin................17, 28, 49, 145, 147
flavone. .35, 67, 70, 198, 200, 204, 209
flavonoids....................36, 200, 207
flavoprotein..............................123
flaxseed.....................................280
Forkhead...................................240
fosinopril...................................281
gamma-carboxylated..............261
garbage.........................267p., 286
gastrointestinal.........................269
genes....4, 10, 12, 18pp., 24, 34p., 50, 52, 58, 61pp., 65pp., 72, 78p., 81, 91, 106, 108, 111pp., 122p., 127p., 132p., 135p., 138, 140p., 149, 153, 160, 162pp., 168, 172p., 175, 187, 189, 192, 194pp., 211, 213pp., 220, 224, 228, 232, 234, 237, 241pp., 258, 263, 266, 275p., 283pp., 288
genetically11, 57, 72, 94, 113, 123, 131, 280
genistein...............................67, 277
genome...21, 41, 123, 128p., 141,

212, 214pp., 222p., 229, 232, 235, 256
genomics....10, 40, 132, 144, 169, 220, 287
genotoxic..................194, 199, 224
germ....................14p., 72, 76, 269
glioblastoma...........................214
glucagon..................................124
glutamine..................17, 50, 91, 96
glutathione.......................162, 224
glycation..................................268
glycotoxic................................248
glymphatic..........5, 30pp., 175pp., 181p., 184pp.
GTPases.....................................93
guanfacine...............24, 100p., 105
guanosine....................93, 112, 214
gyrus................................200, 275
haematopoiesis........................233
haemodialysis..........................283
haploinsufficiency....................109
haptoglobin.............................195
Hayflick................61p., 72, 88, 270
HDACs.....................................209
heart. 5, 12p., 19, 39, 62p., 69, 96, 120, 125, 131, 133, 145, 152, 157, 171, 216, 236pp., 244, 247pp., 256, 258, 262, 264, 267, 269pp., 276pp., 282pp., 286pp., 313
    cardiac. .39, 129, 147, 152, 154, 173, 224, 234, 237pp., 258, 261, 267, 269, 271, 276, 281, 283, 286
    cardiomyocyte. 222, 234, 239p., 269pp., 286pp.
    cardiovascular. 5, 26, 28, 73, 81, 115, 117, 121, 126p., 142, 145, 154, 160, 164, 172, 184, 190, 249, 253, 256, 261p., 267, 277p., 280pp., 284p., 311
cardiovascularevents.............285
heartbeats..............................248
infarcted...............................287p.
infarction.........88, 146, 149, 272
myoblasts...............194, 198, 288
myocardial....88, 146, 149, 272, 282
myocardium..........................287
hearts....5, 39, 236pp., 240p., 273, 288
hematopoietic. 38p., 62, 199, 221, 224p., 235p., 238, 269, 272
hepatocytes......................136, 144
heterochromatin..........................89
high-calorie..............................280
hippocampus.....36, 57, 94, 100p., 194, 199pp., 205, 208, 288
    hippocampal....101, 104, 199p., 202, 204, 208p., 275, 288
Histology..................................258
histone....27, 35p., 88p., 134, 140, 143, 201pp., 207pp., 211, 241, 243
homeostasis..4, 12p., 17p., 22, 34, 38p., 48, 56, 58, 61, 63, 84p., 95, 108p., 111, 113, 121, 124, 140, 146, 178, 192p., 205, 221, 223, 226, 228, 233p., 236, 238, 244, 247, 249, 253, 257, 261, 263p., 266, 268p., 271, 275p., 279, 287
hormesis....................157, 165, 171
hormones........................251, 285

therapies..35, 43p., 53, 89, 153, 157, 175p., 188, 244, 275, 280, 285
HSCs........38p., 62p., 221, 224pp., 231p., 236
HSPCs...................................225pp.
hTERT.......................................77
human5p., 9pp., 14, 16pp., 26pp., 33, 35pp., 39, 44, 49pp., 62, 67p., 70p., 73p., 77p., 81, 85, 88, 91p., 94, 97pp., 102, 105, 108, 116p., 120pp., 124pp., 133pp., 140, 142pp., 147, 149p., 152, 154pp., 160pp., 164pp., 170pp., 175p., 179, 182pp., 188, 196pp., 209pp., 213, 215p., 218pp., 222, 228, 231pp., 235, 237, 240pp., 244pp., 252, 256pp., 262, 264, 268pp., 276p., 279, 281, 285pp., 310p.
Hutchinson-Gilford....4, 22, 84, 95, 114
hydra...................42, 193, 269, 287
hyperlipidemia...........114, 279, 284
hypertension 28, 40, 55, 92, 145p., 152, 279pp., 286
hypertrophy 224, 234, 237pp., 242, 267, 286
hypomethylation........................277
ibuprofen...................................262
immortality......................................
 immortal......................72, 269p.
immortalization.........................270
immune system 78, 196p., 256, 266
immune-mediated..............81, 262
immunoglobulin..........................52
immunophilin...........22, 84, 87, 107
immunoregulation......................283
immunosuppression.....................92
immunosurveillance....................78
indomethacin.....................198, 200
infection.....75, 92p., 96, 111, 115, 127, 249, 255p., 274
inflammation. 13, 58, 65, 125, 132, 163, 172, 194, 196pp., 200, 252p., 255, 261pp., 266, 276, 284
 inflammatory11, 30, 125, 145p., 151, 155, 158pp., 163pp., 170, 173, 194, 200, 255, 261pp., 266p., 279, 284
inhibitor...23p., 27, 36, 43, 49, 51, 76, 78, 85, 90, 92, 94, 96, 100, 104, 106, 112pp., 119, 134, 140, 143, 152p., 181, 187p., 195, 199, 201, 203pp., 208p., 211, 232, 241pp., 258, 264, 280pp., 286
 inhibition....13, 20, 22pp., 29p., 34, 49, 52, 67, 69, 84p., 87, 91pp., 96, 104, 106p., 109, 111p., 114, 119, 122p., 128, 143, 155, 158, 181, 192, 196, 200, 204, 252, 263, 281pp.
injuries..............................251, 269
insulin.......24, 26p., 51, 54, 57, 65, 69p., 75, 88, 106, 108p., 111, 113p., 117, 119, 124p., 128p., 132, 134pp., 139, 141p., 144, 161p., 166p., 174, 176, 263pp., 279
Interleukins...............................284
inulin...181
invasive.....................183, 185, 256
invertebrates....11, 16, 108p., 117,

119p., 156pp., 211, 223, 257, 261
iontophoresis..............................100
iPS........198, 222, 240, 272pp., 283
iPSCs...222
ischemia.....................................271
islet...........................59, 242, 269
isocitrate....................................224
isoflavones............................67, 70
isomerase.....................................52
keratinocytes.................78, 80, 232
ketamine....................................179
kidney. .114, 120, 157, 162, 214p., 258, 264
kinases..................67, 162, 176, 267
knock-in....................................284
knockout. 19, 65, 72, 88, 109, 111, 119, 225p., 241, 267
l-carnitine................28, 149, 152pp.
lamin............23, 28, 84, 88, 90, 95, 145pp., 154, 202
Latrodectus................................120
LDLs...258
lentiviral.............................227, 272
lentiviruses................................226
leptin...................................54, 138
leucine....................................239p.
leukotrienes...............................165
levamisole...................................50
libitum....26, 116p., 122, 124, 128, 134p., 139
life-saving..................................255
life-shortening...........................276
life-threatening....................93, 249
lifespan......6, 9, 12, 16, 18, 22, 26, 28p., 39, 42, 56p., 59, 88, 94, 113, 131p., 143, 171, 233, 235, 246pp., 258, 261, 264, 266, 268p., 276p., 280, 285p.
lifetime.....68, 133, 248p., 271, 277
limbs......................................244p.
lineage-committed....................287
lipids 17, 61, 133, 142, 146, 163p., 174, 239, 280
lipolysis.....................................122
lipopolysaccharide....................198
lipoprotein......125, 136, 140, 146, 163, 175, 186, 190, 253, 280pp.
lipoproteins...............................146
lipotoxic....................................248
longest-living.....................248, 285
longevity...2, 4, 6, 16p., 21, 23pp., 27, 29, 37, 48, 52, 57, 59, 63, 66, 68, 74, 81, 88, 92, 94p., 106pp., 111, 113p., 117, 119, 122, 124p., 128p., 131pp., 135, 138, 140, 143, 155, 157p., 160, 170pp., 184, 193, 208, 210, 212, 214, 223, 231, 233, 235, 237, 241, 246pp., 258, 263pp., 267p., 276p., 283, 285p.
lungs..........................74, 258, 276
lymphocyte....................77, 81, 261
lymphoma.......80, 92, 96, 114, 264
lymphopenia.............................127
machinery............................44, 252
macrolide................22, 84, 87, 107
macromolecules.................205, 264
    macromolecular............252, 268
macular degeneration...............313
Maintenance 14, 22, 43, 72, 79, 81, 84, 93, 97, 100, 175, 187p., 202, 216p., 244, 258, 263pp., 268pp., 275pp., 286

mammalian...4, 22, 24, 27, 29, 37, 39, 49, 61, 68, 72, 84p., 87, 95p., 106, 114, 119, 134p., 142, 155, 167, 202, 210p., 223, 232, 234, 236pp., 248, 257, 261, 269, 276, 287
mannose...................................146
marrow-derived........................272
mechanistically..........................285
medflies....................................120
medial...............................261, 279
melanocyte.......................194, 199
melanogaster..51, 55, 58, 85, 106, 120, 122, 130, 141, 158, 263, 285
mellitus.....................................282
memory......4, 23p., 34pp., 75, 85, 91p., 94, 97pp., 104p., 141, 143, 182, 189p., 195, 197, 199pp., 207pp., 275
menaquinone............................261
mesenchymal....................43, 271p.
meta-analysis..131, 149, 152, 154, 166, 171, 173, 219, 279pp., 284
metabolism 4, 13, 15, 24p., 31, 34, 63, 83, 85, 87, 106p., 114, 116, 119, 122pp., 130, 140, 143p., 146, 148, 154, 158, 165, 171, 176p., 182, 192p., 211, 223, 233, 235, 247, 256p., 261
   metabolic...16, 31, 34, 68p., 73, 80, 96, 114, 119, 122, 131, 142p., 147, 156, 173, 177, 188, 192, 211, 218, 223p., 233, 248, 263, 276, 280, 284, 311
Metabolomic......................149, 239
metalloproteinase......................176

metazoan....................................85
metformin 20, 23p., 27, 41, 67, 69, 85, 93p., 96, 106, 112, 115, 128, 132, 134, 141p., 144, 217, 231, 235, 280
Methodol..........184, 220, 253, 272
Methylation........36p., 88p., 201p., 210pp., 228, 235, 277
methylomes...............214, 216, 220
metronidazole...........................149
microarray.................132, 138, 144
microbiome..5, 28, 145pp., 150pp.
microbiota.........................150, 154
microenvironments....................273
microglobulin............................195
middle-aged.....98, 127, 174, 176, 187p., 202p., 287
mimetics.......27p., 51, 132, 134p., 141p., 144, 264, 279
misfolding.............................49, 56
mitochondria. 4, 13, 16, 18pp., 50, 57, 60pp., 73, 80, 114, 123, 126p., 131pp., 156p., 160, 162, 164p., 170pp., 223pp., 231, 233pp., 241p., 265, 267, 280, 283, 285p.
mitochondrial-directed.............267
mitohormesis.............157, 165, 171
mitokine.....................................64
mitomycin...................................89
models..11, 16, 23, 35, 37, 44, 55, 61, 84, 90pp., 96, 144, 150, 156, 164, 173, 197, 200p., 205, 210p., 215, 231, 241, 261, 283
modules...................................255
   modular........................252, 255
   modularity...........................255

molds..................................288
monoamine..........................104
monocytes.................146, 154, 283
mononucleated.........................271
mortalin..................................57
mortality.14p., 33, 40, 42, 72, 127, 130, 132p., 153, 171, 184p., 190, 244, 277p., 285, 287
mTOR..22pp., 84p., 87, 91pp., 96, 106pp., 111pp., 119p., 124, 126, 128, 132, 135, 138, 167, 264
multicellular.14, 120, 193, 216, 246
muscle......13, 26, 29p., 34, 39, 50, 65pp., 74p., 108, 122p., 127pp., 132p., 139pp., 146, 152, 155, 160pp., 166pp., 170, 172pp., 194, 199, 222, 224, 229, 232, 235pp., 240, 242, 247, 269, 271, 273, 286
mutant. .22, 52, 57p., 68, 72p., 91, 95, 226, 265
mutations. 10, 21, 40, 50p., 53, 73, 90, 95, 108p., 119, 124, 138, 226, 229, 255
myeloma..................................21
myoblasts..................194, 198, 288
myocytes........66, 222, 234, 239p., 269pp., 287p.
myokine................................163
myosin............................50, 272
mysticetus..............................277
naked mole-rat.........................285
nano-devices...........................256
nanoparticle..............................55
NAPDH.................................261
natriuretic.............................238
neoblasts....................43, 72, 270
neonatally..............................277
nephrin........................32, 54, 58
nephropathy..........................282
neprilysin..............................176
nerve..........................32, 34, 276
nervous system...............196p., 248
networks 15, 38p., 41, 43, 99, 200, 217, 221pp., 225, 227pp., 232, 238, 275
neu....4p., 11, 22p., 29pp., 39, 49, 56pp., 69p., 75, 84p., 90pp., 97pp., 104p., 114, 123, 127, 152, 169, 174pp., 181p., 184pp., 192, 194pp., 204p., 207pp., 220, 222p., 233, 236pp., 241pp., 269, 271, 275p., 285, 287p., 310p.
neural........5, 33p., 98p., 189, 192, 195p., 199p., 209, 223, 233, 288
neuregulin...........................271, 287
neurodegeneration.....92, 127, 184
neurogenesis...........34p., 91, 192, 194pp., 237, 241pp., 275p., 288
neurons..23, 31, 39, 69, 75, 97pp., 123, 176p., 195, 199, 201, 204, 222, 238, 269, 275, 288
neuroplasticity..........................209
neuroprotective........................58p.
neurotrophic.....35p., 200pp., 207, 209, 288
neurotrophin..........35, 200pp., 208
Neurotrophins........................208
NF-kappaB.......................172, 280
nicotinamide...........66, 223p., 261
nitrogen....................................264
non-ACE..................................282
Nonsteroidal....................262, 279

NSAID..............................262, 285
NSAIDs............................262, 285
nutraceutical.....10, 20, 36, 55, 66, 207, 241, 256, 263, 268, 279
nutrients....87, 107, 119, 122, 171, 252
obesity.......................133, 140, 284
olive..280
organisms 11, 25p., 28, 30, 42, 51, 72, 85, 87, 106, 116p., 121, 134p., 142, 145, 147, 150, 192p., 211, 216, 222p., 244pp., 264p., 268p.
organs. 10, 31, 43, 52, 62, 77, 170, 176, 244pp., 251p., 267, 269pp., 273p., 283
osteoblast....................75, 141, 144
osteocalcin................................261
osteopontin.......222, 232, 236, 242
Overexpression 57, 65, 69, 87, 112, 171, 227, 233, 237, 241p., 277, 283, 285
Overweight........................131, 284
oxidase..162, 224, 261, 266p., 286
oxidation 65p., 122, 132, 140, 147, 160p., 164, 168, 267, 281
   oxidase-mediated..................286
   oxidized.................146, 222, 266
oxide......28, 38, 119, 145pp., 151, 154, 160, 162p., 165, 221, 224, 227p., 261, 265, 286
oxygen 13, 16, 28, 38, 61, 69, 126, 138, 145, 147, 155p., 163p., 171, 174, 221p., 232p., 235p., 263p.
p-selectin....................................173
pancreas............................242, 282
pancreatic..............241p., 269, 287

parabiosis..............39, 195, 237pp.
paraquat.....................................265
parenchyma.....31, 178p., 189, 249
parenchymal..............................249
pathogenesis. .149, 153, 175, 187, 189, 284
pathogens.........................256, 266
pathological..40, 50, 90, 149, 176, 249
pathology. .31, 58, 90, 150, 175p., 283
pathophysiology..................95, 257
pathway.......11, 13, 18, 21, 32, 36, 50pp., 55p., 66, 71p., 76, 78, 85, 91, 104, 106, 112, 115, 117, 119, 121, 124, 126, 135p., 138p., 144, 147, 153, 157, 167, 172, 178, 189, 201, 205, 207p., 234p., 252, 255pp., 262pp., 266p., 270, 276, 280, 282, 286, 313
penicillin............................249, 278
perinatal......................................275
perindopril..................................281
peripheral....32, 92, 128, 141, 147, 178, 283
permeability...............................267
peroxisome.....27, 62, 69, 73, 134, 139, 143, 162, 224, 234
PGC.....19p., 62p., 65pp., 73, 114, 162pp., 173, 234p., 280
phagocytosis..............................176
pharmaceutical.....18, 56, 77, 256, 263, 271, 274, 276, 310p.
phenylbutyrate............................54
phenylephrine........................239p.
phosphatidylcholine..........147, 154

phosphodiesterase......................104
phosphoenolpyruvate. 24, 106, 108
phosphoinositide......108, 119, 263
phosphorylation...13, 57, 67, 69p., 81, 87, 107p., 136, 162p., 204, 208
phytochemicals...................122, 281
planaria...42p., 72, 193, 269p., 287
   planarian 43, 72, 193, 269p., 287
plasticityn.....................................204
platelet...................12p., 146, 261p.
pleiotropic..................261, 279, 284
pluripotency.............................232
pluripotent 43, 218, 222, 229, 240, 266, 269p., 272, 287p.
pollutants....................................269
polyglutamine............17, 50, 91, 96
polyphenol....................55, 58, 158
populations.....169, 193, 212, 219, 225, 236, 241, 255
post-mitotic 36p., 51, 77, 193, 210, 212
PPAR-gamma...............................62
PPARalpha..................................143
PPARdelta...........................144, 279
prediabetes..........................40, 280
pressure-lowering......................281
primates....27, 121, 123, 132, 134, 142, 157, 264, 281
pro-healthspan..........................256
pro-inflammatory..............158, 165
pro-oxidants...............................168
pro-survival..........................87, 288
progenitor 193p., 196, 199p., 223, 225, 232p., 242, 261, 268pp., 277, 288

progeria 4, 22, 28, 84, 88, 95, 114, 145, 202
progerin...................22p., 84, 88pp.
proliferation...19p., 51, 66, 71, 89, 127, 163, 195, 199p., 223p., 227, 233, 266, 268p., 271, 286pp.
prophylactic....10, 40p., 101, 255p.
prostacyclin............................163p.
prostaglandins...........................165
proteases....................................176
proteasome..........................48, 52
protectin.............................159, 163
proteins. 17p., 20, 32p., 44, 48pp., 57, 61pp., 65p., 88, 119, 123, 126, 151, 162, 175p., 178, 211, 218, 222pp., 246, 261, 268, 273, 277, 287
proteolysis...................................48
proteome....................................41
proteostabilins.............18, 53, 55p.
proteostasis..........................48pp.
PUFAs....30, 155, 158, 160, 167pp.
pulmonary........73, 81, 248p., 253, 257p., 262, 264, 275, 278, 283p.
pyrroloquinoline....................67, 70
quercetin...............................67, 70
quinone........................67, 70, 261
racemization.................................
   racemization..................210, 277
   racemized............................248
radiation....................170, 198, 285
ramipril......................................282
rapalogs.....................................112
rapamycin...4, 22pp., 37, 51p., 57, 84p., 87pp., 106pp., 111pp., 119, 128, 131p., 135, 142, 167, 210,

217, 264, 280
re-engineering............................248
receptor....23p., 27, 33pp., 44, 50, 62, 65, 69p., 73, 78, 97pp., 104, 115, 124, 131, 134p., 139, 143p., 146, 151p., 162, 168, 176, 181, 188, 192, 198, 201p., 205, 207, 224, 234p., 258, 267, 281pp.
recombinant......144, 209, 287, 310
redox. .22, 67p., 79, 84, 87, 106p., 161, 222p., 225, 228, 233, 266, 286
reductase...........................224, 253
reenervation...............................274
regeneration......................................
   regenerate............................269
   regenerating.............42, 72, 275
   regeneration.....15p., 43p., 130, 223, 232p., 237, 242, 253, 268pp., 277, 287, 313
   regenerative...1p., 9, 16, 41pp., 72, 225, 227, 244, 247p., 263, 268, 270pp., 275p., 286, 313
regenerative biology..............41pp.
regression..........................213, 279
regulatory....15, 30, 37pp., 44, 69, 155, 167, 201, 205, 210, 221p., 225, 228p., 238, 283
rejuvenative..................14, 193, 268
remodeling....201p., 208, 223, 281
renin............................257p., 282p.
renin-angiotensin.......................283
reperfusion........................266, 271
replicative. 19, 61p., 66pp., 72, 89, 198, 202, 270, 277
reproductive......................246, 268

reprogramming 222, 229, 235, 273, 288
   reprogram 9, 222, 229, 232, 235, 272p., 288
reptiles...277
respiration......61, 66, 138, 171, 283
resveratrol 23, 29, 67, 69, 85, 93p., 96, 112p., 115, 142, 155, 158, 163p., 171pp., 217, 223, 231, 235, 264, 280
retroviral.....................................272
reversine.....................................271
revolutionize.........................43, 276
rheumatoid................174, 262, 284
ribofuranosyl-imidazole.....27, 134, 141
ribosyltransferases......................223
rifampicin................................17, 49
RNAi.............................50, 122pp.
rodents..
   mice. .11p., 16, 19pp., 31, 34p., 38p., 43, 51p., 55pp., 62p., 65, 67p., 70pp., 84, 87p., 90p., 93pp., 101, 108p., 111, 113p., 116p., 119p., 127p., 131p., 134, 136, 138pp., 147p., 150, 157p., 163p., 173, 175, 179pp., 184, 189, 192, 194pp., 200, 208p., 211, 218p., 221, 223pp., 233pp., 248, 258, 263pp., 267, 270, 276, 280pp., 287
   rats. .11, 25, 31, 49, 56, 58, 100, 108, 116p., 120, 127p., 131pp., 141, 143, 164, 173, 182, 189, 200, 202pp., 209, 258, 265, 279, 281, 283, 286

rodent.........27, 32, 52, 91, 120, 123pp., 134, 140, 158, 164, 179, 184, 201, 216, 258, 261, 265, 273, 283, 285
ROS-based........................157, 267
ROS-mediated............................61
s-Loba.............................79, 218
salicylic...............................262
sarcopenia.........................65, 68
scaffold..........................273, 288
scar....................69, 95, 123, 273
scavengers.....................224, 264
schema................................247
secretagogues.......................278
secretory.............................269
self-assembly.......................274
self-duplication....................287
senescence..........................
   senescence......19, 34, 42, 61p., 67p., 72, 76, 81, 89, 132, 173, 192p., 198, 200, 202, 224, 227, 232, 247, 267, 270, 286p.
   senescent..........19, 62, 211, 286
SENS.....12, 16, 22, 24, 27, 29, 38, 49pp., 54, 58, 61p., 65p., 70, 72, 75, 84, 87pp., 106pp., 119, 128p., 134, 136, 141p., 144, 155p., 161p., 169, 196, 221, 223, 225, 227pp., 232, 234, 247p., 264p., 273, 275
serum......125, 131, 133, 136, 146, 161, 163, 174, 194, 196, 198, 262
Signaling18, 21pp., 27, 29, 32, 35, 38, 51, 57, 63, 67, 71, 76p., 82, 84p., 87, 94pp., 99pp., 104, 107, 109, 111, 113, 115, 121, 131, 134pp., 138p., 143, 155, 157, 161, 165, 167, 172, 175, 181p., 194, 199, 201pp., 207pp., 221, 223, 225, 228p., 232, 240, 258, 266p., 279p., 282, 286p.
silicon-inorganic......................274
sinensis.................................59
siRNA..............................44, 90
sirolimus...................92, 96, 114
sirtuins.............69, 158, 223, 233
small-molecule....77, 153, 244, 287
smoking....................217, 248, 253
sod.....27, 38, 54, 100p., 119, 134, 140p., 143, 160, 162pp., 203, 221, 224, 226pp., 231, 234, 265, 286
soma...18pp., 37p., 43, 48, 53, 55, 58, 61p., 68, 71p., 76, 79, 81, 87, 90p., 95, 107, 119, 129, 176, 210p., 213, 217, 221p., 232, 263, 268pp., 272, 277, 279
spatial....98p., 102, 104, 195, 209, 275
statin.................................
   statin-mediated....................279
statins...............................
   statin 113, 152, 203, 213, 240p., 243, 253, 261p., 278p., 284
stilbene................................55
Streptococcal........................249
Stress.....16, 19, 22, 38, 50, 52pp., 57pp., 63p., 67, 69, 81, 84, 87, 107, 117, 119, 121, 123, 131p., 156p., 161, 163, 166, 171, 174, 176, 189, 194, 199, 205, 211pp., 220pp., 225pp., 231pp., 238, 248, 255, 261, 263pp., 267, 277,

284pp.
subventricular..................194, 275
superoxide 38, 119, 160, 162, 221, 224, 227p., 265, 286
supplements..................................
   supplement 5, 10, 18, 21, 28pp., 34, 37p., 40, 67, 92, 132, 148, 152pp., 158, 161, 165pp., 173p., 197, 217, 221, 228, 231, 256, 261, 264p., 275, 278, 283, 285
   supplementation...5, 28pp., 34, 67, 132, 148, 152pp., 161, 165pp., 173p., 197, 256, 261, 265, 278, 283, 285
surgery..
   surgery.............9p., 40, 274, 288
   surgical......39, 195, 237pp., 256
synthetic biology........9, 41pp., 313
systems biology.....................41pp.
targeted...18p., 44, 115, 123, 242, 283, 285
taurine..54
telomerase........4, 19pp., 61p., 65, 71pp., 76pp.
telomeres..4, 19pp., 60pp., 71pp., 77p., 81, 218
   telomere..........4, 15, 19pp., 36, 60pp., 68, 71pp., 76pp., 129, 211, 218, 286
   telomere-dependent.............286
TERT-mediated...........................76
testosterone.....................252, 278
TGF-beta.....34, 192, 194, 222, 236
thioflavin...............................17, 49
threonine...........22, 66, 84, 87, 106

thrombin....................................176
thrombocytopenia.......................92
thrombogenic...........................261
thrombomodulin.......................173
thromboxane..............................13
thymocytes.................................80
tissue-engineered....................288
tissue-specific........36, 76, 219, 269
toxicity 17p., 55, 58p., 91, 96, 171, 266
trachea.....................................286
tradeoffs.......11pp., 26, 30, 155p., 160p., 167p., 170, 255
trajectory..................................246
transcription 44, 48, 55, 62, 65, 76, 87, 89, 95, 107, 112, 114, 119, 123, 136, 159, 176, 201, 203, 208, 215, 223p., 228, 234, 238, 240p., 268, 272, 286
transcriptome.............................41
transgenic 27, 51, 58, 62p., 65, 80, 90, 134, 136, 140, 189, 195, 267
transition............................91, 267
translation 22, 44, 48, 84, 87p., 90, 107, 119, 142, 208, 263
transmembrane..................54, 261
transplants.....................................
   Transplant......91, 93, 96, 114p., 225p., 253, 273pp., 283
   transplantable......................274
   transplantation..91, 93, 96, 225, 253, 273p.
   Transplanted........................273
transport.....63, 68, 123, 146, 151, 176, 181, 224, 266
transthyretin.......................54p., 58

trauma....................................209, 287
treatment..13, 17, 23p., 30, 35pp., 49p., 75pp., 85, 89pp., 93p., 106pp., 112pp., 119, 123, 128, 132, 139, 141, 144, 147pp., 156, 161p., 169, 188, 198, 201, 203pp., 207, 217, 227, 229, 231, 237, 240, 247, 273, 280pp., 284
triglyceridemia......92, 96, 111, 140
triphosphatases............................93
trkB................35p., 201pp., 207pp.
twistedness................................180
ubiquitin..........................48, 69, 90
ucMGP.....................................283p.
uncarboxylated........................283
uncoupling................................286
vaccination...........................9, 255
vagal...................................248, 276
ventricle........................152, 274p.
verapamil....................................55
vitamins...............29, 155, 163, 265
warfarin-induced......................279
whale............................7, 248, 277
woman................................14, 248
worm...11, 18, 48pp., 79, 87, 119, 158, 223, 268
wound-healing.........................266
Xenopus............................223, 233
  free radicals................17, 172, 264

About the Authors

James W. Larrick MD PhD

Founder, Managing Director,
Scientific Director
Panorama Research Institute
1230 Bordeaux Drive
Sunnyvale, CA 94089
    and
Managing Director, Chief Medical Officer
Velocity Pharmaceutical Development
400 Oyster Point Boulevard
South San Francisco, CA 94080

Dr. James Larrick is a biomedical entrepreneur with an international reputation in biotechnology [cytokines, therapeutic antibodies, molecular biology, pharmaceutical drug development] having written or co-authored eight books, over 250 papers/chapters and >forty patents in his twenty-five year career. He has served on the editorial board of six journals. Dr. Larrick's work on therapeutic antibodies and other protein therapeutics has spanned the whole range of biopharmaceutical product development from target discovery, process science to clinical trials.

Dr. Larrick received his MD and PhD. degrees from Duke University School of Medicine, Durham, NC as a Medical Scientist Training Program scholar. After house-staff training in the Department of Medicine at Stanford University School of Medicine, he completed a post-doctoral fellowship in the Stanford Cancer Biology Research Labs working on therapeutic human monoclonal antibodies for cancer and infectious diseases. In 1982, he continued this work as a founding scientist of Cetus Immune Research Labs, Palo Alto, CA, where he became Director of Research in 1986. While at Cetus he pioneered the use of PCR for the construction of recombinant antibodies. This technology was critical to the development of antibody library cloning and the practical development of recombinant antibodies as a new class of biotherapeutics.

In 1991, Dr. Larrick co-founded PIMM, now called the Panorama Institute of Molecular Medicine, a non-profit research institute situated near Stanford University and Panorama Research Inc., a biopharmaceutical incubator company. Dr. Larrick's PRI team has discovered and initiated development of a diverse, and innovative portfolio of pharmaceutical molecules addressing major unmet needs in cancer, infectious, autoimmune, cardiovascular, neurological and metabolic diseases. PRI has incubated >30 life science projects. Based on this work he has co-founded more than a dozen companies. Among these are Planet Biotechnology Inc., Kalobios Inc., NuGen Technology Inc., Panolife Products Inc., PanResearch Inc., Adamas Inc., Absalus Inc. (now Teva Inc.), Humanyx Ltd., TransTarget Inc., Larix Bioscience LLC, and Galaxy Biotech LLC. Two companies have been co-founded in Europe, PanGenetics b.v. and TargetQuest b.v. To date PRI-initiated projects and/or companies have led to six IPOs/exits. Currently Dr. Larrick serves on the Boards of several early stage companies. Recent work at PRI has focused on Applied Healthspan Engineering--the utilization of advances in molecular medicine to preserve well-being as we age.

Since 1998 Dr. Larrick has led the biopharma screening committees of various Bay Area angel investment groups, most recently serving on the Board of Life Science Angels. He currently serves as a Managing Director and Chief Medical Officer of Velocity Pharmaceutical Development LLC, based in South San Francisco.

Dr. Larrick has organized and led a number of biomedical expeditions, including studies of nutrition, malaria, genetics, and high altitude adaptation among native peoples of Ecuador, Peru, Guatemala, Nepal, India, Tibet and China. He has a genuine interest in fostering entrepreneurial activities and promoting healthcare among those less fortunate. Presently he helps fund and serves on the Boards of two non-profits, the Sustainable Sciences Institute (www.ssilink.org) focused on education and delivery of appropriate technology to less developed countries in Africa and Latin America and the Sankofa Center for African Dance and Culture focused on education, diagnosis and therapy of HIV/AIDS and tuberculosis in Ghana (www.thesankofacenter.org).

Andrew Mendelsohn, PhD

Director of Research
Regenerative Sciences Institute
Sunnyvale, CA 94089
amend@regensci.org
     and
Director of Molecular Biology at the
Panorama Research Institute
Sunnyvale, CA 94089

Dr. Andrew Mendelsohn has been deeply interested in the biology of aging for many years. He founded the 501(c)(3) non-profit Regenerative Sciences Institute in 1994 to pursue research at the interface of aging, regeneration, and what is now called synthetic biology. In 1997, he co-authored one of the earliest papers in synthetic biology. He serves as Director of Molecular Biology at the Panorama Research Institute, which seeks to develop state-of-the-art therapeutics and is a co-founder of Wintgen LLC which seeks to alter Wnt pathway regulation to cure macular degeneration. He serves as Director of Research at Regenerative Sciences Institute (www.regensci.org), which seeks to overcome aging by using synthetic and computational biology to engineer enhanced regeneration and rejuvenation. Development of appropriate dynamic computer simulations of aging biological systems is ongoing. At the heart of our approach is the creation and insertion of new biological programs into cells to augment pre-existing incomplete regeneration mechanisms.

www.ingramcontent.com/pod-product-compliance
Lightning Source LLC
Chambersburg PA
CBHW080345300426
44110CB00019B/2505